The Loneliest Polar Bear

The Loneliest Polar Bear

*

A True Story of Survival and Peril
on the Edge of a Warming World

KALE WILLIAMS

CROWN
NEW YORK

Published in the United States by Crown, an imprint of Random House,
a division of Penguin Random House LLC, New York.

CROWN and the Crown colophon are registered trademarks of
Penguin Random House LLC.

The author owes a great debt of thanks to Oregonian Media Group,
which published "The Loneliest Polar Bear" in its original form in October 2017.

LIBRARY OF CONGRESS CATALOGING-IN-PUBLICATION DATA
Names: Williams, Kale, author.
Title: The loneliest Polar bear / Kale Williams.
Description: First edition. | New York : Crown, [2021] |
Includes bibliographical references and index.
Identifiers: LCCN 2020049765 (print) | LCCN 2020049766
(ebook) | ISBN 9781984826336 (hardcover) |
ISBN 9781984826343 (ebook)
Subjects: LCSH: Polar bear—Ohio—Biography. | Polar bear—
Infancy. | Human-animal relationships. | Zoos—Ohio—Powell.
Classification: LCC SF408.6.P64 W55 2021 (print) |
LCC SF408.6.P64 (ebook) | DDC 599.78609771—dc23
LC record available at https://lccn.loc.gov/2020049765
LC ebook record available at https://lccn.loc.gov/2020049766

Printed in the United States of America on acid-free paper

crownpublishing.com

2 4 6 8 9 7 5 3 1

First Edition

*Title page art from an original photograph by
Dave Killen © Oregonian Media Group.*

For Kale Alonzo Williams Jr. We miss you, Grandpa.

Contents

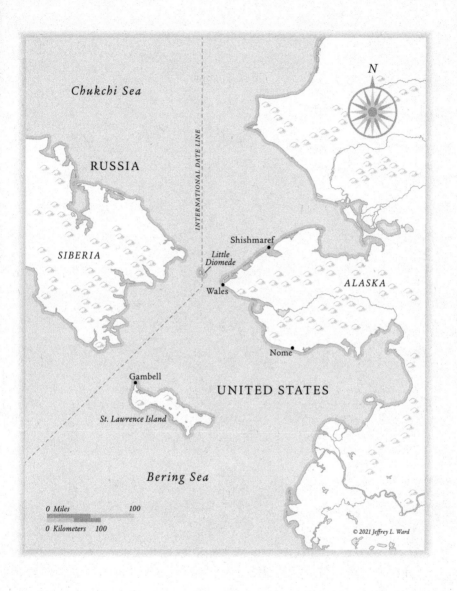

The Loneliest Polar Bear

Chapter 1

Abandoned

She weighed scarcely more than a pound, roughly the size of a squirrel. Her eyes and ears were fused shut. Her only sense of the world around her came from smell, and her nose led her in one direction: toward the gravity and heat of her mother, a six-hundred-pound polar bear named Aurora.

Their den was made of cinder block, painted white and illuminated by a single red bulb in the ceiling. The floor was piled high with straw. The air, heavy with captive musk and kept artificially cool to mimic the Arctic, was pierced periodically by the cries of Nora, a pink-and-white wriggling ball of polar bear, tucked into the folds of her mother's fur.

The tiny cub slept a lot, waking only to nurse, which she did greedily and often, with a soft whir that sounded like a tiny outboard motor. She suckled even in her sleep, her curled tongue lapping at the air.

Around nine o'clock on the morning of Nora's sixth day, Aurora rose, stretched, and ambled out of the den. The cub was completely reliant on her mom, alone and vulnerable without her. As the chilly air crept in around her, Nora cast her head from side to side, screeching as she searched for something familiar, something warm. When she found no answer to her cries, she began to wail.

Outside the denning compound, three women monitored what was happening. Zoo veterinarian Priya Bapodra peered at a grainy, red video—a live feed from inside the polar bear den—as a pixelated Nora squirmed on the screen in front of her. Zookeeper Devon Sabo took notes. Carrie Pratt, a curator, looked on. For five days, the women had worked in rotating shifts, keeping a twenty-four-hour watch on Nora, craning their necks to discern what was happening on the video monitors and pressing headphones to their ears, listening for any signs of distress.

When Nora was born, on November 6, 2015, she was the first polar bear cub to live more than a few days at the Columbus Zoo and Aquarium, which had opened in 1927. The den where she spent her first days was nothing like where she would have been raised in the wild, but it was as close as humans could muster in the suburbs of central Ohio. Nora's birth in that concrete den represented all the ways humans and polar bears were inextricably tangled—for better and for worse. To some, Nora would become the wild north made approachable, an ambassador for a species few would ever see in the wild. To others, she was the physical embodiment of the political battle over whether humans were causing irreparable harm to the planet, a question settled by science long before her birth. Whether she liked it or not, she and her species had become the sad-eyed face of climate change. She represented the damage humans had done to the earth, and she offered the thinnest hope of setting things right.

But to the keepers in the trailer, she was not an ambassador or a symbol. Nora was a helpless cub who was in peril.

And so, at 8:55 A.M., as Aurora took one step away from Nora and then another, the women steeled their nerves and tried to stay calm. Aurora had left Nora alone before, but only for brief periods. In the wild, a mother polar bear never leaves the den, even to eat. The eight-year-old mother wandered down a hallway, past the food her keepers had left for her, and toward the other side of the enclosure. Sabo made a note in the log:

"Aurora gets up and goes into pool room."

Soon after, phones around the zoo buzzed. An alert went out over a text message thread to the rest of the animal care team, letting them know something was amiss. Ten minutes passed. Maternal instincts are innate in animals, but Aurora appeared conflicted.

Bapodra kept an eye on the clock. Twenty minutes now.

As the time ticked by, the tension in the trailer grew. Nora's cries reminded the keepers of their own children, only louder and more urgent. As long as her vocals were strong, they were willing to wait.

Most polar bear cubs born in captivity live less than a month. Only about a third survive to adulthood. When keepers are forced to raise the cubs themselves, the odds are worse. Cubs can't regulate their temperature on their own. Without their mothers, they succumb to disease and infection. They suffer from malnutrition and bone issues because their mother's milk is impossible to replicate. The keepers knew all that when they created Aurora's birth plan, drafted long before she went into labor. The twenty-three-page document was kept in a binder in the denning compound, and each member of the team had a copy on their phone. The plan accounted for all conceivable scenarios, including pulling a cub from its mother. "It will not be possible to return the cubs to the female when their condition improves or they have been stabilized, as she will not accept them," the plan read.

The women in the trailer knew that if they stepped in to help Nora, there would be no going back. The responsibility of raising the helpless cub would fall to them. Between them, the women had decades of experience hand-raising jungle cats, live-stock, and primates. But none of them had ever raised a polar bear. There were only a handful of people in the world who had even tried.

At the one-hour mark, something had to be done. Sabo went into the compound, carrying more straw to coax the wandering mom back to her cub. She walked along the narrow path called the keepers' alley and quietly dropped the straw next to the den where Nora lay crying.

Aurora didn't respond.

Another hour went by and Sabo went into the denning com-pound again. This time she brought fish. On the text thread, Sabo relayed what was happening. Soon, other keepers showed up to watch. Questions swirled in their heads. Could something have driven Aurora from the den? What else could they do to encourage her to return? How long should they wait?

Three hours had gone by, and now the keepers gave Aurora a deadline: one more hour. If Nora appeared to weaken, they would swoop in sooner. None of them wanted to raise Nora themselves. Her odds would plummet the instant they plucked her from the den. But they didn't want to stand by and watch her die, either. Left alone, her odds were zero. They grabbed a plastic bin and lined it with heated water bottles and blankets. Without her mother's warmth, Nora had to be getting cold.

At 12:43 P.M., almost four hours after Aurora left the den, Nora's cries weakened ever so slightly, and she looked sluggish. It was November 12, Bapodra's birthday, and the veterinarian had plans with her husband that night. She called and told him to put the plans on hold.

It was time.

* * *

A little less than half a million years before Nora was born, the earth was going through a significant warming period.

It was hot enough that part of the Antarctic ice sheet collapsed and sea levels rose dramatically, in some places more than sixty feet higher than they are today. Sea surface temperatures rose, too, and coral flourished in balmy shallow ocean waters. The ice sheet that covers Greenland receded, and boreal forests took root on the shores of the usually frigid island. A band of brown bears migrated north, taking up residence in these forests and in similar places around the Arctic, colonizing northern latitudes normally off-limits to them.

And then, roughly 400,000 years ago, the climate changed again. Temperatures dropped, glaciers re-formed, and the northern brown bears were cut off from their southern counterparts. The evolutionary tree split and a new species was born: the polar bear.

At least that's one theory. Others place the polar bear's divergence from the brown bear (known colloquially as the grizzly) at four to five million years ago. Some say it happened more recently, within the past 200,000 to 300,000 years. Polar bears are still so closely related to brown bears that it's possible for them to interbreed, and they have done so many times over their evolutionary arc, especially when a warm climate has allowed the two species' ranges to overlap. (Recent documentation of these interbred bears birthed the term "pizzly," a polar bear–grizzly hybrid that sports a beige coat.)

That interbreeding, the mixing of DNA from the two species, is one reason the exact timing of the evolutionary split remains elusive, and there are few ancient polar bear fossils from which to draw conclusions. However long they've been around, though, polar bears must have managed to rapidly specialize in order to survive in the harsh conditions of the Arctic.

Over generations, as forests gave way to snowy tundra, lighter-colored brown bears found themselves with better camouflage than their darker cousins. As a result, they were better hunters and reproduced more successfully, passing their lighter-furred genetic code on to their offspring until the entire species was white. Eventually, polar bears would have two layers of fur: an outer coat of clear hairs with hollow cores, which provide excellent insulation, and a dense inner layer of shorter hair. Their ears shrank to conserve heat. Their paws grew larger to provide better traction on snow and ice, and webbing formed between their toes to help them swim. Fat grew in a thick layer beneath their skin and enshrouded their organs, to fight off the bitter cold.

Their diets changed, too. Where the brown bear mostly ate grass, fruit, insects, and the occasional fish, the polar bear's diet would come to consist almost entirely of blubber-laden animals from the ocean. Polar bears are opportunistic hunters, eating whatever they can catch easily, including walrus and small whales. But they have historically relied on seals for food. Seal blubber was the only food dense enough in calories and plentiful enough in Arctic waters to support an animal that can grow to 1,700 pounds, up to half of which can be fat. And so polar bears came to bring the full measure of their cunning, patience, and brute strength to hunt their main prey. Their sense of smell became so powerful that they could locate seals through feet of ice. They learned to locate the air holes seals use to breathe, and wait. They learned that sometimes that wait could last for days. When the seal surfaced, the bear learned to plunge its long neck into the water, haul out the seal, and crush its skull.

A diet so rich in fat would prove fatal to humans, but polar bears persist, even thrive, with extremely high levels of cholesterol. As their hair changed color and their paws got bigger, their genes changed, too. Polar bears gained the ability to metabolize vast amounts of fat without clogging their arteries, taking it out of the bloodstream and storing it for insulation. But to get the fat they need, polar bears need seals. To get seals, they need ice.

Since they diverged from their brown cousins, polar bears have ranged across the Arctic, from the icy shores of the islands in the Canadian Arctic Archipelago to the far reaches of northern Russia to the frozen coast of Alaska. As those places see ice cover diminish, polar bears have less access to their main food source, and nutrition plays an integral role in reproduction. Mother bears generally give birth in December and January, and new moms spend the next several months rearing their cubs in dens before they're ready for exposure to the outside world. Adult bears mature late, have few litters, and expend a great amount of time and energy raising their young. Without proper nutrition, females lose weight and give birth to fewer, weaker cubs. Sickly cubs are less likely to survive, so as ice disappears, so too does the polar bear.

Given the threats their wild counterparts are facing, zoos are particularly invested in the survival of captive bears. Nora was one of only two surviving cubs born in the United States in 2015. A twin brother had lived less than two days and died with an empty stomach. Zookeepers believe he never tasted his mother's milk.

Every cub—wild or captive—shoulders a share of the burden of a species in peril. That burden weighed heavily on Nora and her keepers as they tried to figure out what to do with the abandoned cub.

Devon Sabo was the first to enter the polar bear compound. Nora was alone in the den, bathed in red light. Aurora, her mom, had left the small room and wandered toward another room on the other side of the denning compound. The keepers, who had been watching a live video feed from a trailer next to the building, knew this was their chance to act.

Sabo went to the far side of the denning compound with a plate of smelt, one of Aurora's favorite snacks. She grabbed a fish with a pair of tongs and called Aurora. She was distracting the bear so Aurora wouldn't notice as the door slid shut behind her.

Curator Carrie Pratt came in next. Moving slowly, she quietly secured the door with a padlock. As it clicked into place, any remaining bond between Nora and her mother was severed.

Chapter 2

A Fateful Hunt

Nearly twenty-eight years before Nora's birth and 3,615 miles to the northwest, Gene Rex Agnaboogok set out from his house on the edge of the Inupiat village of Wales, the westernmost community on the North American continent.

He packed only what he needed: warm clothes, a rifle, cigarettes, and coffee, as well as extra gas and food in case he got stuck. It was late in March, but the temperature was still well below zero, and sea ice stretched far out into the Bering Strait. Heading out alone, he knew there would be no one to help him if something went wrong.

He gunned the engine on his snowmobile and cruised along the waterfront, past the remnants of traditional houses built of sod and newly assembled constructions of wood and steel, past the tall cross sticking out of the snow-covered dunes that marked the spot where hundreds of his ancestors were buried. Heading northeast, Agnaboogok was soon in untouched country. On his right, a vast lagoon sat frozen and unmoving. To his left, in the middle of the strait, the Diomede Islands marked the International Date Line and the border between the United States and Russia. It was a clear day, and beyond the islands, Agnaboogok caught glimpses of the Siberian coast.

Three hours into the hunt, Agnaboogok had come up empty.

The stereotypes of the north—cold, empty, and barren—were proving correct. Agnaboogok hadn't seen more than a passing fox in the distance, too far to get a shot off and too small to warrant chasing down. Then, just before 11 A.M., he found prints so large they could come from only one animal. An adult polar bear could feed several families for weeks. Agnaboogok followed the plodding tracks away from the coastline and out onto the Bering Strait.

Close to shore, four-foot waves of ice sat frozen in time—a rough, unmoving sea. Wind whipped around the cape, picking up snow and piling it in drifts. Agnaboogok's snowmobile bumped along over the ice until the sea stretched flat for hundreds of yards. There, where the ice had been blown smooth, Agnaboogok lost the bear's tracks. He throttled back the engine. Now nearly a mile from shore, he needed to regroup.

He was close to the edge of what is known as fast ice, sheets that are fixed to the mainland, and at the foot of a pressure ridge. Currents in the strait, where the Pacific and Arctic oceans mix and swirl, had stacked icebergs up into towers, three or four stories tall, that stretched in a line parallel to the shore. Beyond the ridge, broken ice and ocean. Agnaboogok picked the tallest iceberg and hiked up to get a better view. He pulled a Marlboro Light from his shirt pocket and sipped coffee from his thermos, scanning the landscape with binoculars for anything he could hunt.

The only discernible movement: waves on the open water and blowing snow.

In the early 1970s, all five nations that are home to polar bears—the United States, Canada, Denmark (which governs Greenland), Norway, and the Soviet Union—joined with the International Union for Conservation of Nature to estimate their global population. Using observations from Alaska Natives and other sources, American researchers said they thought there were about eighteen thousand bears. Canadian experts thought it was closer to twenty thousand. Soviet bear counters thought there could be as few as five thousand bears roaming the northern reaches of the planet.

Despite the uncertainty over which of these population estimates was accurate (if any), there was agreement among the Arctic nations that the animals were being hunted at unsustainable levels and that conservation measures needed to be put in place. In 1973, they formalized that accord in the Agreement on the Conservation of Polar Bears, which drastically limited the circumstances in which the animals could be killed and banned the commercial trade of their skins, bones, and skulls. All of the signatories pledged to devote resources to the study of polar bears. In the midst of the Cold War, the agreement was among the few things upon which Soviets and Americans could find common ground. Canada instituted a quota system and outlawed hunts from aircraft or ships. Norway put a five-year ban in place. The United States banned polar bear hunts altogether, with exceptions for Native subsistence hunters like Agnaboogok.

Around the time those protective measures were put in place, Ian Stirling, who would become an icon in the field, began his research on polar bears. He had grown up in a small mining town in the mountains of western Canada before getting his master's degree in zoology from the University of British Columbia. Stirling had always been an outdoorsman, but one of his earliest forays into polar research came in the late 1960s, when he wrote his Ph.D. thesis on the population dynamics of Antarctica's Weddell seals. In the early seventies, he headed north and landed on the western shore of Hudson Bay, in Churchill, Manitoba. Much of what we know about polar bears was learned in that remote pocket of northern Canada. Hudson Bay is home to one of the world's southernmost populations of bears, and researchers have taken advantage of the bears' relative accessibility. Churchill would serve as Stirling's home base for Arctic research for much of the next five decades.

Studying polar bears is rarely easy or comfortable. The first challenge is finding them. Polar bears are solitary animals that roam vast and shifting ranges. Much of their lives are spent on sea ice, some of which is adrift on the ocean, decoupled from land, and which researchers generally canvass from helicopters. Even then, a white animal in a white environment is not easy to spot.

Early polar bear research aimed primarily to get an accurate count of the animals and to determine whether a given group of bears—there are nineteen distinct subpopulations circling the top of the globe—was increasing, decreasing, or remaining stable. And that was done with rudimentary tagging studies conducted over multiple years.

In the first year of a mark-recapture study, researchers head out in a helicopter, toting rifles loaded with tranquilizer darts. Once they find a bear, hit it with a dart, and watch it drift into sedation, the aircraft sets down and they begin taking measurements of the animal's length and girth. They may set up a scale on a mobile tripod and weigh the bear, collect a biopsy, or take samples of its fur or fecal matter.

A small plastic tag, slightly larger than a human thumbnail and bearing a unique identification number, is affixed to the bear, sometimes one in each ear. Today, ear tags can contain GPS trackers and other advanced monitoring mechanisms, but back when Stirling first started, all they had was numbers. Researchers also use a special tool to tattoo a unique ID number on the inner lip of the bear, in case the ear tag falls out. Then they get back in the helicopter to look for the next bear. That process is repeated for as many bears as can be found during the tagging season in the first year.

Then they wait.

The second year of a mark-recapture study looks a lot like the first one. Researchers fly around in the same area as they did in year one, tranquilizing and tagging as many bears as they can and recording how many were marked with ear tags the year before. Then they calculate the ratio of marked to unmarked bears to estimate the size of the population for that area. If you catch and tag one hundred bears the first year, then catch another hundred bears the second year, 10 percent of which have ear tags, you can assume that about 10 percent of the overall population was tagged in year one. If 100 bears is 10 percent of the population, you've got yourself a population of 1,000 bears.

But the mark-recapture method is far from perfect. It assumes that each animal in a given study area has an equal chance of being captured every year, and that animals don't wander in or out of the study area. Then, too, polar bears of different ages and genders can segregate themselves into different types of habitats. Male bears may prefer open sea ice, while females with cubs may congregate in areas with more cover, so male bears may be over-represented in population estimates. Some areas are too remote for any type of estimate at all.

Given these pitfalls, it should come as no surprise that there is disagreement about the status of many polar bear subpopulations. But what we do know is that while several are classified as stable, only a few of the subpopulations are growing. At least three have seen declines in recent years. In the southern Beaufort Sea, along the northern coast of Alaska and Canada, researchers have seen a 40 percent drop in polar bear numbers in less than a decade. In Canada's western Hudson Bay, where Stirling is based, the population dwindled at least 32 percent since the late 1980s before leveling out. And in southern Hudson Bay, bears are thinner and dying younger. Meanwhile, the total number of polar bears that roam the Arctic remains a mystery. The best guess puts the number somewhere around 26,000.

Long before Agnaboogok clambered to the top of that iceberg in 1988, his Inupiat ancestors and other Natives had hunted in the far north. Fossils from a woolly mammoth, discovered in Siberia, show injuries that could only have come from human weapons, presumably crafted by some of the Arctic's earliest inhabitants. The Dorset people are thought to have settled in what is now the Canadian Arctic by 500 B.C.E. They hunted almost exclusively on the sea ice, using triangular blades made of stone, waiting for seals to appear at air holes and harpooning walrus and whales. They made lamps from soapstone, fueled by seal oil. They had chisel-edged tools called burins, which were probably used for the intricate carvings and complex masks they fashioned. It's not clear exactly what happened to the Dorset, so named for the cape where their remains were first discovered, but their demise, between the years 1000 and 1500, coincided with a shift in climate. Anthropologists believe animal migration routes were likely disrupted and the sea ice the Dorset relied upon changed.

The decline of the Dorset made way for the Thule, who are thought to have settled first in Alaska as early as the first century and then spread out to the east, some going as far as Hudson Bay. Experts believe the Thule used sealskin kayaks to hunt on the open sea, wielding harpoon heads and knives made of polished slate instead of chipped stone.

The Thule were the ancestors of the Canadian Inuit and the Alaskan Inupiat.

Gene Agnaboogok's earliest Inupiat ancestors are thought to have populated Alaska's north and west coasts as early as the year 1000. Theirs was a culture of cooperation and sharing, of trade between villages. Agnaboogok's hometown, which Europeans would later call Wales, was named Kingigin for the mountain that rises behind it. The people of the village—between six and seven hundred at its peak—called themselves Kingikmiut, "the people of Kingigin." They hunted seals, walrus, bowhead whales, and polar bears, like their predecessors had done. Elders in villages like Wales learned the currents that pulled icebergs away from the fast ice attached to shore. They learned the migration patterns that brought different kinds of prey within hunting distance. They learned to harvest what they could from the unforgiving land and sea. Each generation refined that knowledge, over thousands of years, and passed it on to the next. That's how Agnaboogok learned to hunt. That's what brought him to the top of that iceberg, looking fruitlessly for game to bring home.

Cigarette spent, he flicked the butt into the snow and started down the slope, taking a different route than he'd taken on his way up. Halfway down, coffee in hand, he heard the ice crack. His weight felt suspended in air for a split second, and then he was falling. He crashed waist-deep into one of the many cavities beneath the surface. As Agnaboogok struggled to gain his footing, he realized the ground below him was moving.

He had broken through the roof of a den, and he was standing on top of a polar bear.

Chapter 3

First Feeding

Priya Bapodra stared at the thermometer and at the white cub, who was screeching.

The digital screen said simply LOW.

Just as the vet had feared. Nora, who had now been without her mother's warmth for more than four hours, was so cold her body temperature wouldn't register. The cub wriggled and squirmed and screeched against the intrusion, the strange smells, the foreign feel of human hands. Moments earlier, Bapodra had entered the den and nestled Nora into the warm blankets lining the plastic bin before the vet rushed her to the zoo's medical center, where the cub went straight into the intensive care unit. Nora was small enough to fit in the palm of your hand, disoriented and angry. Bapodra knew that if she didn't stabilize and warm the little bear, she would lose her.

Bapodra chose a 22-gauge needle, the second smallest she had, and searched Nora's thigh for the femoral vein so she could draw blood. It was a vein she knew she could hit the first time, even on a creature so small and squirmy.

"Hang on, girl," Bapodra said as the needle broke through Nora's skin. The vet always talked to the animals, but this time the one she was trying to reassure was herself. Of the roughly seventy polar bears living in accredited zoos and wildlife refuges in North America in 2015, only four had been successfully hand-raised from infancy.

The women who became collectively known as the Nora Moms had their twenty-three-page plan, with step-by-step instructions in the subsection "Removal of Cub." They had a manual: *Hand-Rearing Wild and Domestic Mammals*. They had an incubator the size of a double-wide refrigerator. Bapodra set the small-animal compartment to eighty-eight degrees and lined it with clean baby blankets.

They were as ready as they could possibly be, and they were not ready at all.

A few buildings away, the phone rang in the zoo's nutrition center, a squat beige warehouse where a small crew worked to feed the facility's seven thousand animals. Dana Hatcher, who ran the center, picked up the phone and got what seemed like a simple request:

"Can you come over to the animal hospital so we can talk about polar bear formula?"

She headed to a large conference room—joining a group of Nora Moms, curators, and administrators—and explained the recipe she had on hand.

"So, if you pull the cub—"

Someone interrupted her. "Dana, we've already pulled the cub."

They needed the formula within an hour.

At home, Hatcher cooked like a scientist, using Excel spreadsheets and metric weights. She had cooked for lemurs, flamingos, and her seven-year-old. But never a polar bear cub. She felt like she was on one of those harried time challenges on a Food Network cooking show, but with much higher stakes.

Hatcher knew that polar bears were among the most difficult animals to feed. In the wild, they lived almost entirely off ringed seals, whose blubber added richness to polar bear mothers' milk. The zoo had nothing that would accurately mimic seal fat, and Hatcher knew that past attempts at fortifying young polar bears' formula with heavy cream had been unsuccessful; most cubs rejected it, and even those that didn't sometimes ended up needing medical treatment for nutrient deficiencies. Fortunately for Hatcher, a veterinarian from the San Francisco Zoo had managed to study wild polar bear milk just a few years earlier. To get it, scientists in helicopters had to fire tranquilizer darts at mother polar bears on the frozen fjords in the Svalbard archipelago, north of mainland Norway, then hand-milk them on the ice. Knowing the chemical composition of that milk gave Hatcher an advantage: She knew, roughly, what nutrients Nora would need.

Hatcher had developed a recipe for polar bear milk before Nora was born, but she'd never made it before. She started with a can of powdered kitten milk replacer—baby formula for cats— and sifted it so it wouldn't clump. The formula was low in fat, leaving Hatcher room to add calories however she saw fit.

She knew the sugars in the cat formula would be hard on Nora's stomach, because polar bears can't digest cow milk or goat milk. It took a few tries, but she figured out that heating the water to a specific temperature would help break down the lactose. Hatcher also knew that polar bears need lots of taurine to help absorb vitamins, so she crushed taurine tablets with a mortar and pestle and added that to the mixture. Then came the biggest and most important piece of the puzzle: Nora needed fat, and lots of it, to grow. But what kind?

Human and cow's milk contain around 3.5 percent fat. Polar bear milk is more than eight times as rich, with a fat content of upwards of 30 percent. The birth plan recommended using herring oil, but that was available only from Canada. Shipping it across the border would be complicated, and she didn't have time. Nora needed milk immediately. Hatcher considered her options and went with safflower oil.

She poured it all into an industrial blender and hit the button for low. The oil and the creamy formula spun together but separated like old yogurt. Hatcher tried turning the machine to high. The mixture frothed but ended up too thick to drink. After a few tries, Hatcher still couldn't get the consistency right; the big blender was too powerful. She wished she had the Magic Bullet she used to make salsa at home. It was the perfect size, and it had a pulse function.

She sent a staff member on a quick trip to Target.

The smaller mixer proved to be the key. Hatcher heated the water and added the pre-sifted powder. Once those ingredients were blended smooth, she poured in the oil. She hit pulse on the Magic Bullet once. Twice. Three times.

It was ready.

Cindy Cupps was charged with Nora's first feeding.

The mother of Devon Sabo, the keeper who'd been watching from the trailer when Aurora left the den, Cupps was a zoo veteran whose presence calmed the younger keepers. She was serene when others were stressed, tranquil when they were tense. As she headed to the zoo's intensive care unit, she locked eyes with another keeper named Shannon Morarity.

Morarity was known as the crier in the group. Sure enough, when she caught Cupps's glance, her eyes were welling. Cupps looked at Morarity and clenched her fists.

"We can do this," Cupps said. Morarity looked back at her and took a breath.

"Okay."

As Cupps arrived at the intensive care unit and opened the door to the incubator, she marveled at Nora's head, just a little bigger than a golf ball. She scooped her up. Thin white fur covered Nora's back and legs. Her squeals fluctuated between a high-pitched whine and a miniature roar. Her tongue lapped about freely in search of something to suckle. After not feeding for roughly five hours, Nora was famished. Cupps couldn't help but think about the threat facing Nora and her species, the shrinking sea ice in the Arctic and the upward march of global temperatures. The seasoned zookeeper, who'd cared for hundreds of animals over her career, was struck by the moment.

This is unreal, she thought.

Cupps draped a towel across her thigh and slid her hand under Nora's soft belly. She held the cub upright so she wouldn't inhale any of the formula Hatcher had worked so hard to get just right. The keeper tipped the bottle toward the cub, and Nora latched on. She fed so tenaciously that a small milk mustache formed around her mouth. Cupps encouraged her in a soft voice, using a nickname that would stick.

"Good girl, Bean."

The next day, the zoo's public relations staff posted seventy-six seconds of video from one of Nora's first feedings. Viewers saw Nora, eyes still fused shut and dwarfed by the hands that held her, gnaw on Cupps's gloved thumb. They watched Cupps stroke her with a finger. Under normal circumstances, Aurora would have kept Nora in the den for months, nursing her and preparing her for life outside. When Nora was revealed to the world at just six days old, those watching and rewatching the video took in a sight few outside of zoo nurseries ever get to see. Nora became an international phenomenon overnight.

It was a risk to introduce her to the public so soon, though.

The odds of Nora living were still slim, and if she died, there would be questions. Her fan club, which was growing by the hour, would want to know what had happened. Critics would say the zoo had exploited the imperiled cub, using her undeniable adorableness to bolster their bottom line.

Baby animals are moneymakers for zoos, and few newborns are as rare or captivating as polar bear cubs. In the mid-nineties, a mother bear gave birth to two cubs at the Denver Zoo, which were then abandoned and raised by zookeepers. Like Nora, Klondike and Snow were quickly thrust into the public eye. The zoo gave tours where visitors could get up close and personal with the cubs. An airline painted their portrait on the tail of one of its jets. The zoo produced its own video in-house, called "Saving Klondike and Snow," which was sold at local supermarkets alongside oven mitts, hats, and Christmas ornaments emblazoned with the likenesses of the cubs.

The zoo made only three thousand copies of the video at first, unsure of the cubs' marketability. Two years later, more than ninety thousand copies of the video had been purchased, along with another twenty-five thousand copies of a sequel called "Klondike and Snow Growing Up." The cubs raked in more than $300,000 in royalties on merchandise featuring their tiny faces. "The zoo's mission is conservation and education," the zoo's marketing director, Angela Baier, told *The New York Times*. "My mission is to get people through the gate and to educate them." And Klondike and Snow were nothing if not effective at getting people through the gate. Attendance nearly doubled from the year before the cubs were born, and family memberships at the facility went up by a third. Nora presented a similar opportunity to the zoo in Columbus.

* * *

Thousands of people all over the world watched that video of Nora's first feeding. Some of the commenters wished to trade places with Cupps, imagining holding something so rare, feeling the strong tug of her mouth on the bottle, hearing the satisfying swallows that meant Nora now had a chance. Because this was the internet, some critiqued Cupps's feeding technique. Nearly all were struck by how adorable the cub was. As Nora's feedings continued, she gained weight at a steady clip. At least for the moment, Hatcher's frenetic experimentation in the zoo kitchen was working.

Chapter 4

The Bear

Gene Agnaboogok's legs slipped and skidded, trying to find purchase on the ocean of white fur beneath him. He felt like he was running on a waterbed. Adrenaline flooded his nervous system. He was a mile from shore on the frozen Bering Strait and waist-deep in a collapsed polar bear den. If he screamed, no one who would hear him, except the angry bear beneath him. He didn't waste his breath.

Agnaboogok managed to plant the butt of his rifle in the snow. He spun, struggling to free himself. He thrashed, heart hammering. As the bear moved underneath him, he leapt from the hole and threw himself as far as he could onto the ice, hoping the bear would stay in the confines of her den. Behind him, she lunged.

A three-inch claw tore through his snow pants and into his right leg just above the knee. Her nose hit the back of his leg, and her lower jaw struck his heel. Had she turned her head sideways and latched onto his leg, he likely would have died out there on the ice. Instead, the bear's attack launched Agnaboogok out of the collapsed den and onto the side of the iceberg. He scrambled on his hands and knees and turned just in time to see the mother bear come out of the hole in the ice and circle around it uphill from where he sat.

She reared up on her haunches, standing several feet taller than Agnaboogok. He whipped his rifle into firing position and took aim. At a dead sprint, polar bears can cover nearly thirty feet per second. This one was fifteen feet away. Agnaboogok drew a breath and pulled the trigger.

The bullet ripped a hole in the bear's thick hide where her neck met her torso and she crumpled, sliding down the iceberg and back toward the den. She used the last of her energy to draw her body as close as she could to the opening in the ice, leaving a trail of blood down the frozen slope.

Breathless, chest heaving and exhausted, Agnaboogok approached the bear carefully. He thought he'd seen her move. When he was taught to hunt, he had learned to never let an animal suffer needlessly. Most Inupiat hunters would rather pass up a risky shot from far away than risk their prey getting away wounded, only to die a slow and painful death hours or days later. Agnaboogok fired a second shot, just below the skull, to make sure the bear was dead. He didn't have time to thank her or make offerings to her spirit, as was the Inupiat custom, because as he stood over her body, Agnaboogok saw movement inside the den. He jumped back and then leaned forward, peering into the dark. Two sets of small black eyes, framed in white fluff, stared back.

Baby polar bears, motherless on the ice.

It was at that moment that Agnaboogok noticed his pants were warm and wet. Figuring he'd spilled his coffee, he reached under his ski pants. His hand came up red, covered in the blood leaking from the hole in his leg. He was twenty miles from home, nearly a mile out on the sea ice, injured and alone.

The cubs would have to wait.

*   *   *

A few months after Gene killed the bear, Dr. James Hansen, director of NASA's Goddard Institute for Space Studies, walked into a U.S. Senate committee hearing room on Capitol Hill.

It was stiflingly hot in Washington, D.C., that June, just as it was in many parts of the United States as the country grappled with a historic drought. Few areas were left untouched by the lack of precipitation. The Upper Mississippi watershed hadn't seen so little rain since the Dust Bowl. The river itself was low, hampering navigation on the nation's most important waterway and driving up the cost of water for irrigation and consumption. Less water meant less hydroelectricity, just as homeowners and businesses cranked up the air-conditioning to try to fend off the stifling heat. Many utilities had to switch from hydropower to fossil fuels to meet demand.

The corn yield in 1988 was nearly cut in half, and soybeans were down 25 percent. Many grain harvests saw dramatic losses. In the West, nearly seventy thousand wildfires consumed more than six million acres of forestland. Tourism, retail sales, and agricultural income all took huge hits. The extended lack of rain cost the country more than $51 billion. The impact of the drought went beyond the land and the money made off it, though. Rates of asthma also increased, and mental health problems spiked as farmers worried about their land and city dwellers grew anxious over how they'd pay for their air-conditioning. A majority of states needed federal disaster assistance.

That was the backdrop as Hansen walked into the hearing, took his seat, and told lawmakers that the abnormal heat they felt that day would likely soon be the norm. He told them about what scientists were calling the greenhouse effect, that carbon dioxide had built up in the atmosphere to such a level that it was acting like a blanket for the planet, that the climate was changing in ways that would soon be beyond our control, and that we, human beings, were the cause.

He broke his conclusions down into three major points.

"Number one, the earth is warmer in 1988 than at any time in the history of instrumental measurements," he told the committee and the more than a dozen cameras that beamed his testimony to millions across the country.

"Number two, the global warming is now large enough that we can ascribe with a high degree of confidence a cause and effect relationship to the greenhouse effect. And number three, our computer climate simulations indicate that the greenhouse effect is already large enough to begin to affect the probability of extreme events such as summer heat waves."

Hansen didn't discover climate change, or global warming, as he referred to it. Scientists had been aware of the potential effects of a warming climate, and humanity's role in causing it, for more than a hundred years.

On a smaller scale, humans had suspected that regional climate change was possible since ancient times. Theophrastus, a student of Aristotle and "the father of botany," theorized that when lands were stripped of trees, a localized rise in temperatures could be expected. Academics in the Middle Ages thought the extensive grazing, irrigation, and deforestation associated with the spread of agriculture around the Mediterranean were responsible for changes in the weather.

That idea was brought to scale as European settlers turned much of eastern North America from forest to farmland by the nineteenth century. The changes to the natural landscape were so drastic and so swift that they could have been observed over the course of a single lifetime. The agrarian lifestyle swept across North America like a fire, displacing whole societies of Indigenous people, jumping the Mississippi River as it emptied the land of trees and grasses, leaving neat rows of crops in their place. Over the three centuries following European settlement, three to four acres of land were converted from forest to farm for each person added to the population. Those moving west theorized that their work on the ground had effects on the skies. "Rain follows the plow," they told each other. Tilled soil released its moisture to the air, which then fell back as rain, proponents argued. Newly planted crops increased precipitation, too, they said. Others thought that human activity increased vibration in the atmosphere and caused rain clouds to form. They were right that human activity could cause changes to the climate, but it wasn't in the way they thought.

While homesteaders in North America sought to take advantage of how the climate changed, scientists in Europe began to discover the first scientifically reliable signs of large-scale climate variability. For many years, odd geologic formations in Alpine valleys had flummoxed those who sought to understand the natural world. A hunter named Jean-Pierre Perraudin, who lived in the Swiss Alps, saw huge boulders sitting where they shouldn't be, obviously carried in and dropped there by some unseen force. Deep gouges in the solid rock above his home in the Val de Bagnes had no easy explanation. For years, prevailing theories suggested that a great flood, perhaps the one described in the Bible, had moved the titanic stones, or that pressure exploding from the earth's core had launched them into their present positions. But Perraudin knew that rocks didn't float, and he'd seen the glaciers high in the mountains above his home. He knew their power, great enough to carry huge stones or carve marks into valley walls. He figured that at some point in the past, the ice sheets that in the 1800s were limited to high peaks had stretched all the way down to the valleys below.

Perraudin brought his idea to prominent naturalist Jean de

Charpentier, but he was roundly rejected. "I found his hypothesis so extraordinary and even so extravagant that I considered it as not worth examining or even considering," Charpentier later wrote of the meeting. Still, Perraudin persisted. When Ignace Venetz, now seen as the father of glaciology, came to Val de Bagnes, the hunter saw his chance. Venetz was convinced, and the idea that the world had once been much colder worked its way up the chain of scientific influence. Venetz persuaded the once skeptical Charpentier, who in turn convinced Louis Agassiz, then one of the most authoritative voices in the European scientific community. Agassiz was among the first to be credited with the theory of a broader ice age, a period during which ice had once flowed down from the North Pole to cover much of Europe and North America. Like those before him, Agassiz was initially met with skepticism from his contemporaries, but his descriptions of a previously snowbound Northern Hemisphere captured the public mind, and by the 1870s his theories were widely accepted.

The realization that the earth's climate was prone to wide variations gave nineteenth-century scientists a lot to think about. If the earth had existed in a wholly different state in the past, what had caused it to change? Could it happen again? Could humans—and the increasingly complex societies they'd built under relatively stable climate conditions—adapt if it did?

Perhaps the most important question posed in the search for what caused dips in temperatures on a global scale came from Swedish scientist Svante Arrhenius in 1896: "Is the mean temperature of the ground in any way influenced by the presence of the heat-absorbing gases in the atmosphere?"

That question—whether gases in the atmosphere can trap heat and warm the earth—represented a watershed moment. Arrhenius's theory, and his research that bore it out, represented the first time anyone had used the principles of physical chemistry to link an increase in atmospheric carbon dioxide to higher surface temperatures on the earth. Like all scientists, Arrhenius walked in the steps of those who came before him. Eunice Newton Foote, one of the few female scientists working at the time, had published a study in 1856 showing that an increase in atmospheric "carbonic gas" would result in a corresponding increase in temperature. In 1861, a British scientist named John Tyndall announced that the lesser gases in the atmosphere—water vapor, methane, and carbon dioxide, as opposed to oxygen and nitrogen—were responsible for the greenhouse effect and were thus greenhouse gases. An entry in the journal *Nature* from the early 1880s warned of the link between human activity and rising temperatures. From the author, identified as H. A. Phillips:

> According to Prof. Tyndall's researches, hydrogen, marsh gas, and ethylene have the property in a very high degree of absorbing and radiating heat, and so much that a very small proportion, of only say one thousandth part, had very great effect. From this we may conclude that the increasing pollution of the atmosphere will have a marked influence on the climate of the world.

Phillips missed some stuff in his assessment, namely carbon dioxide, but he wasn't wrong about the link between pollution and climate. Over the following years, articles in newspapers across the United States—including *The Philadelphia Inquirer* and *The Kansas City Star*—pontificated on the relationship between human activity and warming temperatures. In an 1883 article in *The New York Times,* an unidentified author put it starkly:

> Every one must perceive that the growth of civilization and the increase of the number of civilized human beings is intimately connected with smoke. The first point of difference between the

beast and the man is that the latter can build a fire. He does build many fires, and the more civilized he becomes the more fires he builds. Now, the process of combustion develops a variety of noxious gases, among which may be particularly mentioned carbonic dioxide, a gas that is produced in large quantities by the combustion of coal. . . . When we reflect that all the gases given off by burning coal enter and contaminate the atmosphere, and that the latter is a constant quantity while the former is steadily increasing, we gain an idea of the danger which threatens us.

But Arrhenius was the first to succinctly marry all the previous efforts together and put solid numbers to the problem. His 1896 study "On the Influence of Carbonic Acid in the Air upon the Temperature of the Ground" surmised that if levels of carbon dioxide in the atmosphere were to drop, temperatures would cool, causing more snow and ice to form at the poles. That snow and ice would reflect heat away from the earth and cause further cooling. Arrhenius had found a plausible cause for ice ages. But the inverse was true as well: If levels of carbon dioxide in the atmosphere were to double, the world could expect to see a global temperature spike of five to six degrees Celsius.

One of Arrhenius's colleagues, Arvid Högbom, began trying to map out sources of carbon, separating the natural from the human-caused. At that point, the Industrial Revolution had been under way for more than a century and smokestacks stood tall, like monuments to production, spewing pollution into the air all across the developed world. Högbom found that natural emissions of carbon—from sources like volcanoes and the respiration process of plants—were roughly comparable to those from human sources, which consisted mainly of burning coal in factories. The rate of increase was slow for human-produced carbon, though, and neither Högbom nor Arrhenius thought humans had the capacity to put enough of the gas into the atmosphere to make a difference for thousands of years. If people ever did cause warming, they thought, the results for humanity would be beneficial.

Throughout the early twentieth century, scientists worked to explain how the climate worked and how it could change so drastically from the frigid ice ages to the temperate climate of the 1900s. Geologists examined layers of clay in lake beds, looking for clues about ancient weather, while others looked at tree rings for signs of a consistent climate cycle, but most claims that carbon dioxide was increasing, that humans were the cause, and that the increase was causing temperatures to rise were met with skepticism—laughed at, ignored, or outright dismissed.

By the time Hansen went before Congress in 1988, climate science had grown exponentially. Investments in research after the Cold War had helped fund advances in digital technology, where vast data sets could be processed quickly to produce reliable climate models. In 1958, Charles Keeling applied a novel method he'd developed for measuring carbon dioxide in the atmosphere and showed that concentrations rose and fell in a yearly cycle as plants took in the gas during growth periods, then released it back into the atmosphere as they shed their foliage. Within a few years of his first observation, Keeling began to see a signal emerging from the noise. Within the seasonal rise and fall of carbon concentrations, there was a sharp upward trend in the longer term. His graph showing the ascending trajectory would come to be known as the Keeling Curve. It represented a breakthrough in climate science and has been hailed as one of the most important discoveries of the twentieth century. The measurements Keeling took in 1958 are still used as the global standard for carbon levels in the atmosphere.

In 1981, Hansen published the first analysis of global temperatures over time, which showed that the planet had warmed by nearly half a degree Celsius between 1880 and 1980. Using Keeling's scale of measurement, Hansen found that carbon levels in the atmosphere had increased from about 300 to 340 parts per million over that same period, mostly from the burning of fossil fuels. He didn't just look backwards, though. Hansen predicted that if carbon dioxide emissions continued to increase unabated, warming would continue in tandem. By the end of the century, the rising temps would go well beyond anything that could be expected from natural variability, and the impacts could be severe. Vast areas of North America and Central Asia would become prone to drought as the climate zones human beings had relied on for centuries shifted. The ice sheet that covers Antarctica would begin to erode, with dire implications for sea levels around the globe. In the Arctic, sea ice would melt away and the great Northwest Passage, the historically icebound shipping route between the Pacific and the Atlantic via the Arctic Ocean, would be open for much of the year. A boon for commercial shipping, but a looming disaster for the people and animals who depend on the sea ice.

Hansen's 1981 paper landed on the front page of *The New York Times,* under the headline "Study Finds Warming Trend That Could Raise Sea Levels." But throughout the early eighties, much of the media's environmental attention, and therefore the public's, was focused on more immediate threats, like nuclear war, acid rain, and the hole in the ozone layer. Even so, conservative pushback on climate science built throughout the first half of the decade. Reagan threatened to cut funding for carbon monitoring and other climate research efforts. A report from the Environmental Protection Agency (EPA) in 1983 warned that catastrophic fallout from climate change could arrive not years from now, but in the near future. We had years to deal with the problem, according to the report, maybe decades—but not centuries. The Reagan administration called the report "alarmist." Climate change was too big, too amorphous, too indistinct to capture much of the public eye.

Hansen's findings were built on the back of hundreds of years of science that preceded him—from Jean-Pierre Perraudin, who wondered how the giant boulders had ended up in his Alpine valley, to Eunice Newton Foote, who theorized that carbon dioxide levels and temperatures rose in tandem, to Charles Keeling, who provided the first consistent measurements of the greenhouse gas. Like stratified layers of bedrock, each of them, and all the scientists in between, added crucial elements to our understanding of how the earth functions. How the climate was shaped by natural systems. How our actions, though seemingly inconsequential in comparison with the planet's size and scale, could profoundly change those same systems we'd come to depend on.

Mentions of global warming in American newspapers jumped after Hansen's testimony, from an average of roughly two per newspaper between 1981 and 1987 to nearly twenty in 1988.

Even then, we knew the path we were on could lead to catastrophe without a change in course. We knew it was of our doing. We knew what we needed to do. Under different circumstances, Hansen's testimony and its requisite call to action could have been a tipping point. The movement to deny the findings of climate scientists like Hansen, then in its infancy, would make sure that it wasn't.

Three months before Hansen testified, Gene Rex Agnaboogok had tangled with an angry mother polar bear and won. But now he was twenty miles from home, with a claw-sized hole in his leg and a choice to make: take care of the two orphaned cubs left in the den or tend to his own problems.

The choice wasn't difficult.

He motored back to Wales on his snow machine and stopped by the village's rudimentary clinic. The wound wasn't severe. A couple of stitches, some antibiotics, and a day's worth of rest were all he needed. The next day, his father, Roland Agnaboogok, and nephew, Tony Phillips, followed the snowmobile tracks back to the iceberg to salvage the hide and meat from the dead bear.

Roland stripped the fur from the animal, cutting the hide in a straight line from groin to chin before butchering the polar bear into manageable chunks. Phillips approached the collapsed den, where he saw the cubs, alive but quiet and still. They were only a few months old—each was the size of a small raccoon—but they'd already grown sharp teeth. Phillips took off a glove and dangled it in front of the first bear as a distraction. When it lunged, he grabbed it by the scruff of its neck and put it in a cardboard box strapped to the back of the snowmobile. He repeated the process with the second bear, lamenting the puncture marks the cubs were putting in his new gloves. With no room left in the box, he tucked the second bear into a backpack, letting its head poke out of an opening in the zipper. On the way back to Wales, the backpack bear wriggled out, and both men had to get off the snowmobile to chase it down.

Back at the family home, Gene and his father fed the cubs warm milk and played with them in the living room. They knew they couldn't keep the animals, and they knew they didn't stand a chance alone on the ice. They called the local representative for the U.S. Fish and Wildlife Service, who would put the cubs on the path to captivity.

The next day, the bears were flown to Nome, the closest town of any size to Wales. *The Nome Nugget,* the town's newspaper, sent a reporter to Wales, and Agnaboogok's story was featured in the March 31 edition, along with pictures of the cubs nestled on a bed of straw in a large dog crate. From Nome, the bears were flown to the zoo in Anchorage, where they were given names. One was called Norton, after the Norton Sound, an inlet from the Bering Sea south of Wales. The other was given the name Nanuq, the Inupiaq word for "polar bear." Nanuq would go on to spend time in nearly half a dozen zoos, in Wisconsin and New York and, eventually, Ohio. About twenty-seven years later, he would have a daughter.

Her name would be Nora.

Chapter 5

Signs of Trouble

Of all the Nora Moms, Priya Bapodra was the most worried. Nora had spent less than a week with her mother in the den before keepers rescued the abandoned cub. The young veterinarian knew that despite all of their collective experience raising zoo animals, they would likely fail with Nora. And failure meant death. In those first days after the world watched the video of Cindy Cupps bottle-feeding Nora, Bapodra's mind raced with questions. Had Nora gotten enough antibodies from her mom to ward off pneumonia? Was she too cold? Was she too warm? Every action came with a round of second-guessing.

Nora's coat was still wispy and thin. The tiny cub, still not much bigger than a squirrel, couldn't regulate her body temperature. In the den, she would have spent most of her time tucked into the folds of her mother's fur. In the zoo's intensive care unit, away from the public-facing part of the facility, the Moms kept her in an incubator called a Snyder box. It looked like an industrial oven in a restaurant kitchen, with double glass doors that swung open from the middle, but it gave the keepers the ability to keep her temperature at a consistent eighty-six degrees. They kept track of oxygen levels in the machine and did their best to replace the feel of Aurora's fur with clean baby blankets.

Nora left the cozy incubator only for meals and medical checks, so she wouldn't get too cold. While she was out, Bapodra counseled the other Moms to finish their tasks quickly and put Nora back.

From the start, Nora had no shortage of willing surrogate caregivers. They kept Nora wrapped tight to their chests, her eyes still sealed shut. She could smell them and feel their heat, though, and Cupps talked to her softly and stroked her head. Shannon Morarity scratched the cub's round belly as Nora tried to bury her face in the crook of the keeper's arm.

The day after she was pulled from the den, Morarity noticed that the pink pads on Nora's feet were beginning to turn black, followed by her nose the next day and the insides of her ears the day after that. Soon the hair that had once barely concealed her pink skin thickened, and she turned whiter and softer. After the first couple of weeks in human care, Nora was putting on weight at the rate of a little more than an ounce a day as she fed from a bottle every few hours on the concoction Dana Hatcher had conjured in the zoo kitchen.

The keepers gradually dialed down the incubator, allowing Nora's body temperature to self-regulate. They kept the ICU dark like a den and tried to avoid talking around Nora so that she could sleep and grow undisturbed. Morarity would stare at Nora in the dark, through the window of the incubator, and watch her roll or snore or sleep—belly up, tongue out, paws twitching as she dreamed.

She would ask herself: *What does this baby need?*

Each of the Moms had a job at the zoo outside of caring for Nora, and those responsibilities didn't end with the arrival of the baby bear. Morarity's shifts sometimes lasted twenty-eight hours. Through the holidays, she barely saw her family. Three weeks after Nora was born, Bapodra ate Thanksgiving dinner on the floor of the ICU, near Nora's incubator. Their families understood. They could tell how much Nora meant to them.

As the days turned into weeks, Nora grew feistier. Her eyes were still closed, but she began moving around the incubator on her own. Her head was too big and her tummy too round for her to stand, but that didn't stop her from trying. The Moms allowed her more freedom, sitting with her on rubber mats on the floor so she could learn to walk. At the thirty-day mark, in early December, Nora's odds of survival, once so remote, improved to fifty-fifty. A coin flip. The Moms breathed a little easier.

Four days later, Nora opened her eyes for the first time. She saw Bapodra, the analytical one, anticipating every contingency. She saw Morarity, the tenderhearted one, who was often crying. She saw Cupps, the veteran, telling the others it was going to be all right. No one can say what she made of the scene before her, but there were no bears in it, and no snow or ice.

Off the west coast of Greenland sits Devon Island, a massive piece of desolate land jutting out of the northwest corner of Baffin Bay in Canada's Arctic Archipelago. On the eastern end of the oblong landmass, an ice cap rises to more than six thousand feet. A small herd of musk oxen lives on the island's protected northeastern corner, known as the Truelove Lowland area, where a microclimate allows for plants to grow and a surprisingly vibrant ecosystem of lemmings and weasels and Arctic foxes to thrive. The island is otherwise a frigid desert—barren, rocky, and mostly lifeless. At more than twenty thousand square miles, it's nearly as big as West Virginia, but it has never hosted sustained human settlement. A few groups have tried—fur trappers and

Inuit hunters—but none lasted longer than a few years. The Arctic desert landscape is so alien to life that NASA uses it to train astronauts as an analogue for Mars.

Ian Stirling's mark-recapture studies had given him insight into the population dynamics of polar bears, but much about the species was still unknown. What did polar bears do when they weren't being chased by a helicopter? How would one behave when it was going about the business of simply being a bear? The only way to answer those questions was to watch the creatures, for extended periods of time, without their knowing they were being watched.

That's what Stirling set out to do when he traveled to Devon Island in the summer of 1973.

He and his team set up a camp at the top of Caswell Tower, a 650-foot-tall headland that gave sweeping views of Radstock Bay below. To the south, Lancaster Sound had already thawed, but the bay was still frozen over. Bears in the area were forced onto the remaining ice in the bay to hunt directly below Stirling's cliffside perch. The outpost was spartan, consisting of an observation cabin and an unheated tent they used for sleeping in shifts. Armed with binoculars, powerful spotting scopes, and folding lawn chairs, the research crew had unimpeded views of bears for miles in every direction, twenty-four hours a day—that far north, the sun didn't set. Unless a bear disappeared behind an ice ridge or was obscured by fog or blowing snow, Stirling or a member of his crew was able to observe everything it did. That first year on Devon Island, the crew observed and recorded more than six hundred hours of bear behavior. They watched as the bears hunted fat ringed seals by lying in wait next to breathing holes where the pinnipeds periodically surfaced. They watched as the bears tried, usually unsuccessfully, to stalk the seals in the water, swimming stealthily toward their prey as they basked on the ice. They watched mother bears tend to their cubs, teaching them to wait patiently, without moving, next to breathing holes.

"I wanted to just let the bears show us, at their own speed, what it meant to be a wild undisturbed bear on the sea in the Arctic," he later wrote of that first expedition.

Stirling would return to Radstock Bay several times over his long career, but perhaps his most important observation came in 1997, when he and his team watched an entire mating sequence between two bears over the course of almost two weeks, something no one had ever witnessed in the wild.

Polar bears have one of the lowest reproductive rates in the animal kingdom. Males can mate by age six, but they face fierce competition from older bears, and many males won't win the chance to breed until they're ten or older. Females reach sexual maturity late, between the ages of four and six, and spend years rearing cubs after they're born. With a life span of roughly twenty years in the wild, a mother bear will likely have the opportunity to give birth to only five litters in her lifetime. On rare occasions, as many as four cubs can be born at once, and solo births aren't uncommon, but most cubs are born in sets of twins.

Polar bears are solitary, but the act of breeding entails elaborate social rituals. When spring comes and temperatures in the Arctic begin to slowly rise, males begin looking for females. Because sows spend so much time rearing their young, only about a third of adult females are even available to breed at any given time, and finding them is no easy feat. It's still unclear exactly how males find potential mates on the vast stretches of lonely ice they inhabit, but the leading theory is that females leave a scent trail of hormones excreted by glands on their feet that lets males know they are able to breed.

Sometime before 3 A.M. on May 2, 1997, on the frozen ice of

Barrow Strait off the south coast of Devon Island, a male came across a female's tracks that likely held the breeding scent. Despite the hour, the Arctic sun still shone brightly as he traced her path to the northwest, directly into Stirling's line of sight. The male bear was three miles from the scientist's lookout, about a third of a mile behind the female, who traveled with a cub that looked to be about two and a half years old. Most females, especially those with cubs, would flee at the very sight of a male bear. Males are aggressive and have been known to kill and eat cubs. On rare occasions, they have been known to prey on full-grown females as well. But this one didn't appear to be hunting. He plodded behind the female, trudging through drifts of snow that had piled up on pressure ridges that gave texture to the otherwise flat ice. The male seemed to be in no particular hurry, and neither did the female or her cub. She knew he was there— she was checking behind her every few minutes—but she didn't run. The cub was more nervous, looking over its shoulder almost constantly and staying so close to its mom that at times, through the powerful scopes Stirling was using to watch them, they appeared to be one bear.

The female and her cub walked steadily, changing directions a few times, but the male circled behind them, herding them toward the protection and privacy of Radstock Bay. When they stopped, he did, too. When the mother and cub lay down to rest, the male followed suit, plopping down in the snow but always keeping his head up and his eyes on the female. Periodically, both male and female would stop and rear back on their hind legs to stand. They looked at each other across the ice, each presumably sizing up the other.

They kept up this slow ambulatory dance for nearly twelve hours, until they crossed the pressure ridges that formed at the mouth of the bay, where the male began to creep closer to the female and her wary cub. He was watching her intently and corralling her, running parallel or even sometimes ahead, all the time keeping her in an area just north of Stirling's observation hut.

Then, around 9 P.M., more than eighteen hours after he began following her, the male made his move.

He closed to within sixty-five feet of the female, and she turned and ran at him, lunging at the much bigger animal. She didn't make contact, though, and all three bears ran together for a short distance before they abruptly stopped. He ran at her this time, positioning himself between the female and her cub. If he was going to kill the young bear, this would have been the time. The cub stood its ground and watched for nearly a full minute before turning to run. But the older bear gave no chase, and the female showed no signs that she cared to reunite with the cub or defend it from the male. Stirling didn't know it then, but he'd just seen mother and cub part ways for good. The young bear followed the bears at an increasing distance for a few days and then it abruptly turned and left. The cub would have to fend for itself from that point on.

The female turned and continued walking, with the male again following about a hundred feet behind her. About forty-five minutes after the cub was out of the picture, the true pursuit began. Physical contact. The bears bit each other around the head and neck, though never hard enough to draw blood. They stood on their hind legs and opened their mouths wide in pantomime aggression.

For days the male kept close, sniffing her tracks as she walked and trotted and ran ahead of him. Their interactions came abruptly and with no predictable pattern. One second they'd be marching along, and the next she would run at him, stopping just short of a collision. She might stand and push him in the chest with her paw or they might lock jaws. At certain points, it looked like the bears were vocalizing, but Stirling was too far away to hear them. Sometimes she would break into a sprint and he would chase, only for the race to end suddenly when she lay down on the ice, where he would join her. All of these behaviors were interludes between long periods of walking, the female always in the lead. When they weren't walking or wooing each other, they slept side by side, close but never touching.

Polar bears are aggressive animals by nature, and males vastly outweigh females. The risk to her was very real, and he had to gain her trust. But there was also a physiological reason for the long courtship. Female polar bears don't ovulate on a regular cycle, like many mammals. If they produced an egg every month and there was no male nearby to fertilize it, the chance at new life, along with precious energy, would be wasted. Instead, females have what is known as induced ovulation. She will produce an egg only after she's spent sufficient time with a male, interacting and allowing him to gain her trust.

For the two bears Stirling was watching in 1997, that took nearly a whole week. On May 9, seven days after he'd first laid eyes on the bears, they began to mate. For five days they copulated frequently, the male mounting the female from behind, sometimes for just minutes at a time, once for more than two hours. They swung their heads from side to side, and she sometimes nibbled at his forelimbs. On a few occasions, the male would rest his head on the back of her neck.

After their initial mating, the female grew submissive, and all signs of aggression in both bears stopped completely. They weren't alone on the ice, though—three other males approached during the mating period. In all three cases, the breeding male dispatched the interlopers, two of them after dramatic physical confrontations. Both of those bears left with blood streaming from wounds in their necks.

After what would prove to be their final round of copulation, a marathon 150-minute session, the female appeared to lose interest. She sniffed at the ice and dug her nose down into seal lairs beneath the snow. The male still tried to herd her, but less vigorously than he had over the previous week, and the distance between them grew. Almost exactly thirteen days after they had wandered into Stirling's sights, they headed north, into rough ice and out of view.

After two bears part ways, the male immediately begins sniffing the snow again, searching for tracks that hold the telltale scent of a female ready to breed.

For the female, the next few months would not be so simple.

If the bear Stirling watched was lucky enough to get pregnant, the fertilized egg wouldn't start to grow in earnest for several months. In her uterus, the cells that would one day become a bear would divide only a few times before going dormant. Polar bears mate in the warmth of spring and early summer, but an expectant mother's body isn't ready to grow a baby until she's built up enough fat reserves to spend months in a den without eating. The female from Radstock Bay would have spent the summer feeding on seals, doing everything she could to tell her body she was prepared. Around October, if she didn't have enough fat stored up, the embryo would either be shed or absorbed into her body. If she'd been successful in her hunting, the egg would implant in her uterine walls and the fetus would begin to grow.

As fall turns to winter, a pregnant bear chooses a site to build her den. Near Hudson Bay, bears build dens in raised peat soil or along the beds of rivers. In other places, they excavate their birthing chambers in snowdrifts along coastal bluffs or far out on the frozen sea. Some bears will den as far as thirty miles inland, others on ice more than a hundred miles from shore. In Alaska,

Nora's grandmother built her den in the cavity of an iceberg in sea ice twenty miles northeast of Wales. Wherever a mother chooses, and in whatever medium, the dens themselves are similar. The female digs a tunnel with a chamber just large enough to allow her to turn around, and often a second room. Then she waits for the winter snows to cover over the entrance, providing insulation from the bitter cold. Once she enters the den, her heartbeat slows by almost half, and she won't eat, drink, or defecate until she emerges in the spring. By December or January, the long breeding process—from the first encounter with a male to the lengthy trust-building exercises to the summer of hunting to build up her fat reserves—will bear fruit. She'll give birth to between one and four tiny, helpless cubs, devoid of eyesight, fur, and teeth. For months she'll nurse the baby bears, until they've grown strong enough to survive outside the protection of the den. She knows to do this from instinct built over thousands of years of trial and error, of evolutionary adaptation to an unforgiving environment.

At the Columbus Zoo in 2015, the women caring for Nora could not rely on instinct built over generations. When Shannon Morarity asked herself, *What does this baby need?* she had no easy answers. Nora never got the luxury of an attentive mother and a protective den. And around her fourth week, that fact would begin to show itself.

While the Moms were busy raising Nora, Dana Hatcher had been in the nutrition center doing research on how to import herring oil, which, according to all the literature, had a fatty acid profile more similar to polar bear milk than the safflower oil she'd been using. Once she'd found a supplier in Canada, Hatcher began to slowly incorporate the new fat into Nora's formula. But just as she began introducing the herring oil, Nora's weight gain slowed. In late December, Bapodra noticed that Nora was passing oil in her stool.

The cub was fifty-four days old now, and Bapodra ordered a full exam. The results were not encouraging. Nora's vitamin D and calcium levels were low. Critical building blocks for growth and bone development, which would have been absorbed easily from her mother's milk, seemed to be passing right through her system.

Bapodra also took X-rays.

In the darkened radiology room just outside the ICU, the vet looked at the images. Nora's brain, still in the early stages of development, showed white on the radiographs. Her spinal column snaked away from her head, as it was supposed to. But when Bapodra looked at Nora's limbs, her face dropped. Morarity, standing just outside the doorway, saw the look in the veterinarian's eyes. She knew something was wrong.

The X-rays showed Nora's bones curving where they should have been straight. She had a fracture in her left radius, one of the load-bearing bones between the wrist and the elbow.

Despite everything Hatcher had done, Nora had not gotten the nutrients she needed and had developed metabolic bone disease.

Bapodra stepped out of the X-ray room and headed to the ICU. Morarity was already there, lying on the floor next to Nora, who was still groggy from the anesthesia. The cub was crabby and hungry and let out a high-pitched growl at her keepers. When Bapodra walked in, Morarity read everything on her face.

This time, both women cried.

Chapter 6

When Death Came by Dogsled

In July of 1776, the explorer Captain James Cook set out from the south coast of England on his third voyage, which would also prove to be his last. Cook had made some of the first maps of Newfoundland and explored numerous islands in the South Pacific. As he approached his fiftieth birthday, Cook had been contemplating retirement. But, lured by the promise of a £20,000 prize to whoever made its discovery, Cook went in search of the Northwest Passage, an elusive trading route from the Atlantic to the Pacific over the top of North America.

His journey took him below Cape Agulhas, at Africa's southernmost tip, and then east, below Australia, before he and his crew bisected the islands of New Zealand and turned back north, navigating the HMS *Resolution* toward the Hawaiian Islands, where they stopped before heading for North America.

Much of the Pacific Northwest, and pretty much everything above it, were still largely uncharted by European explorers, and Cook left a lasting mark on the lands he encountered for the first time. Oregon's Cape Foulweather was named by Cook, for reasons that are clear if you visit, as was Cook Inlet, which stretches from the Gulf of Alaska to the state's largest population center, Anchorage. On his way to the Bering Strait, in search of the Northwest Passage, Cook named Norton Sound after Sir

Fletcher Norton, then Speaker of the British House of Commons.

On August 8, 1778, as Cook sailed north from Norton Sound, he found himself between two continents. On one side was the east coast of Siberia. On the other, continental North America's westernmost tip. "We thought we saw some people up on the coast; and probably we were not mistaken," Cook wrote. He called it the Cape Prince of Wales. Cook never found the Northwest Passage, turned back by walls of sea ice almost four hundred miles north of the cape, and he was killed by Native Hawaiians during a stopover there on his journey home to England. But many of the names he gave to inlets and capes and the piece of land that jutted out into the Bering Strait, just miles from the Asian continent, stuck. On maps, Gene's hometown—Nora's ancestral homeland—is still called Wales, though many of its inhabitants use the Inupiaq name, Kingigin.

Wales was once one of the largest and most prosperous of the more than a dozen villages on the northwestern Alaskan coast, which were bound by common tradition and culture but each retained the independence of a nation-state. The village on the cape was home to a population of up to seven hundred people at a time, split between two, or sometimes three, smaller settlements, separated by less than a quarter mile but each with its own distinct identity and customs.

Most other communities in northwest Alaska had to move frequently in search of food on the unforgiving landscape, keeping winter and summer camps depending on where game was most prevalent. But villagers in Wales had easy access to both the land and the sea, including the migration routes for sea mammals just offshore in the Bering Strait. Wales was one of only two villages in the region that were populated year-round.

The people who lived on the cape hunted caribou, the hides

of which made up much of their clothing. They hunted bow-head whales in skin-covered boats called *umiat,* each animal providing not just several tons of meat and blubber but also baleen, which they used for snares, and bones big enough to build with. Walrus skins provided boat covers, and their tusks were used as tools and weapons. Their most important prey, though, was the bearded seal, called *ugruk.* From the seals they got meat and fat to eat, but they also used the skins for boots and tent covers, the oil to heat and provide light for their homes. When hunters landed a whale or an *ugruk,* the meat was divided up among the village residents, the largest portions going to the elders.

When marine mammals were scarce, they hunted seabirds or fished or gathered clams and crabs. Berries lined the shores of the lagoon north of the village, and they picked willow leaves, preserving them in seal oil. The most successful hunters with the biggest *umiat* (usually the patriarchs of the biggest families) were known as *umialgich.* European explorers called them chiefs, but there was no official hierarchy that awarded them that title. Still, *umialgich* acted as leaders in the community, guiding hunting expeditions, organizing feasts, and resolving disputes in the absence of a formal legal system. As the point closest to Siberia, Wales became a trading post for villages around the region, sending emissaries to nearby communities and reaping the benefits of their good fortune.

Most people in Wales lived in sod houses, one-room dwellings dug halfway into the tundra, with driftwood or whale ribs to support the turf-covered roof. Entryway tunnels, dug lower than the house itself, formed a sort of airlock to trap the heat provided by seal oil lamps, and benches served as sleeping platforms. There was little room inside, so villagers kept many of their possessions outside on driftwood racks that held fishing gear, skin boats, and harpoons alongside fish, meat, and hides.

Their lives followed the seasons. They stuck close to home in the cold and dark winter months, visiting with family and friends. As the sun rose higher in the sky, the men strayed farther on hunting expeditions, often returning only briefly to drop off game for the women to process. Every September, the village hosted a massive festival, drawing visitors from communities around the region for feasts, dancing, and games.

At the center of social life for hunters was the *qargi*, a structure two to three times larger than the average home, built by the village's most successful *umialgich*. The *qargi* served as community center, courthouse, church, dance hall, and boat-building workshop and was usually restricted to use by the men. The *qargi* was also where oral histories, stories of hunting and storms and survival, legends of creation, and tales of animal spirits were passed from one generation to the next. The elders were master craftsmen and hunters, and the *qargi* was where they shared their knowledge with the young.

The sea and land provided plenty, and though there were hard times and conflict just like in any small community, the people of Wales were able to provide for themselves to such a level that the population remained robust for hundreds of years before Russian hunters began making incursions into Alaska in the late 1700s. Then, around the middle of the nineteenth century, the first American whaling ship, a three-masted barge called the *Superior*, sailed into the Bering Sea. Its crew was astounded by the number of bowheads that populated the cold waters off Alaska's west coast. The following year, a rush ensued, and during the peak season in 1852, nearly 2,200 whales were killed. Between 1849 and 1857, roughly a third of the Arctic's bowhead population was harvested.

Foreigners on whaling ships also introduced the Native population to alcohol and all the problems that came with it. The people of Wales were wary of the outsiders, who often stumbled off their ships drunk and tried to seduce the women of the village. Native hunters, plied with alcohol, were coerced into making terribly uneven trades, sometimes swapping fur and pelts worth thousands for firearms that cost no more than a few dollars.

In 1877, the trading brig *William H. Allen,* helmed by George Gilley, was anchored offshore when Natives from the cape approached in *umiat* and a fight broke out over alcohol. Some said Gilley ordered his men to fire upon the Natives with repeating rifles, others that the people of Wales attacked unprovoked; either way, the massacre that followed is undisputed. More than a dozen Native men were killed, and possibly as many as thirty. They were buried in a mass grave on the hill behind the village, marked with whale bones.

Wales gained a reputation as a lawless place as the village elders began shunning visitors. Whaling ships avoided the community, having heard rumors of the violent people on the cape. That reputation may have been accurate, but the people of Wales had their reasons to be hostile to outsiders. One visitor wrote that the villagers were trying their best to "protect their people from the debauchment of liquor, swindlers, and worse, the seduction or even rape of their women and daughters. . . . [They were] unable to differentiate in advance between responsible and trustworthy crews and those imbued with evil intent, and so sought to escape all."

Wales persisted even as some other villages began to go hungry. Whales, seals, and walrus were being harvested by ships at high rates, and famished Natives, according to some accounts, took to eating the carcasses of decomposing whales that washed up on the beach. Venereal disease and influenza, imported on the same ships that pillaged the sea, spread in many communities.

In 1877, about a decade after the United States bought Alaska from Russia and the same year as the Gilley massacre, Sheldon Jackson, a Presbyterian missionary, went north under the guise of assisting the Native population of Alaska. Jackson had taught in mission schools in the Choctaw Territory during the mid-1800s and made no secret of his disdain for Native people.

By 1889, acting as the territory's general agent for education, Jackson decided the federal government needed to step in more aggressively to help the Native population.

For Jackson, help meant stripping the Native population of Alaska of their traditional language and culture. He traveled the country to rally support for his mission by speaking of the backward ways of the northern Natives, telling embellished tales and outright lies about cannibalism, polygamy, infanticide, and sexual slavery in Alaska's villages. He interspersed his speeches with promises of untapped resources. Alaska was ripe for exploitation, he told eager crowds, if only the people there could be subjugated and "civilized." To do so, Jackson said, he would establish schools in remote communities, first to teach, then to preach. Beyond the cultural shift he had in mind, Jackson imagined a whole new way of life for the Natives of northwest Alaska. He imagined an Alaska where, supplied with docile reindeer from Siberia, the Native population could be converted from hunters to herders.

Jackson personally oversaw the opening of more than a dozen schools across Alaska, and his lobbying led to dozens more. Some of them were boarding schools to which Native children were forcibly sent, where lessons were taught only in English and the use of Native languages was punished. Curriculums were heavily laden with biblical themes. The mission school in Wales, founded in 1890, educated generations of children in English and Christianity.

Jackson was largely successful in his push to acculturate the people of Alaska to European norms. Many adopted the missionaries' religion, and use of Native languages dwindled as English-only schools spread. But the colonial mindset Jackson brought with him, which dictated that the natural world was a resource waiting to be converted to capital, never took hold. The Natives treated the natural world as an interconnected system, with each component—the whales, the seals, the ice, the polar bears, and the people—playing a part. None existed alone, and if one part was out of balance, the whole system suffered. Despite the introduction of reindeer herds, subsistence hunting was still an integral part of daily life. In Wales, the population remained steady, game from the land and sea supporting a village of more than five hundred.

Until the winter of 1918, when death came to Wales on a dogsled.

No one knows exactly where the global Spanish flu pandemic started, but the earliest reports of the illness were in Haskell County, Kansas.

Haskell County, in the southwest corner of the state, was farming country. In 1918, its 578 square miles were home to roughly 1,500 people, nearly all of whom worked in agriculture, growing grain, raising cows and chickens, and maintaining a large population of hogs. They were also squarely beneath the migration route for more than a dozen species of birds. Today we know that influenza in birds can also infect hogs and humans and that when it does, it can mutate into spectacularly virulent forms of the disease. We don't know for sure that's what happened in 1918, but it's as good a guess as any.

In January of that year, amid an outbreak of what the local

paper then called pneumonia, several men from Haskell County reported for duty at nearby Camp Funston, a sprawling Army training center housing thousands of recruits from around the region. By mid-March, more than a thousand men were sick, lined up in rows of cots in the camp's hospital ward. From Kansas the illness spread to other camps, ultimately infecting two-thirds of Army training facilities nationwide. When soldiers from those camps were sent to Europe, the true devastation began.

More than 10,300 sailors in the British Grand Fleet were admitted for treatment in May and June, but only four died. Dismissed as a "three-day fever," the illness got little attention from the press, especially in warring countries, where reports that might indicate weakness were either censored or never written at all. When influenza arrived in Spain, King Alfonso XIII became infected, along with the prime minister and several members of the cabinet. The Spanish had remained neutral as war raged around them, and the illness made daily headlines in newspapers across the country. Elsewhere—because the United States, England, and France were afraid to report its presence—the influenza of 1918 became known as the Spanish flu.

Most strains of influenza spread rapidly because they infect the upper respiratory tract—the nose and throat. The Spanish flu was different: The virus embedded deeper in the lungs, which could cause both bacterial and viral pneumonia. While the early version of the 1918 flu killed comparatively few people, those it did kill provided clues about what was to come. In a typical year, most people who die from influenza are those with weak immune systems: the young, the old, or the already sick. But even in the early months of the epidemic, the Spanish flu was causing otherwise healthy adults to succumb. In any other year, the deaths of so many young people might have been cause for more alarm, but the fog of war masked the numbers. By early summer, a medical bulletin from U.S. Army officers in France said the "epidemic is about at an end . . . and has been throughout of a benign type." British doctors said that any sign of the illness "has completely disappeared."

In August, a new and far deadlier strain cropped up in Switzerland, where a U.S. Navy intelligence officer surmised, in a confidential report, that they were witnessing the return of the "black plague." By early September, soldiers at Camp Devens, an Army training center northwest of Boston, began to fall ill. From there the viral infection moved on to Philadelphia, where it exploded, first in the Navy Yard and then in the civilian population after thousands packed the streets for a military parade. More than twelve thousand would perish within six weeks, including almost eight hundred on a single day. The disease extended its deadly tentacles across the country. In Columbus, Ohio, where Nora would be born almost a century later, thousands were sickened and more than eight hundred died of the disease, even as health officials ordered quarantines and closed down gathering spots.

The first cases in the Pacific Northwest were reported in late September at Camp Lewis, south of Seattle. Officials urged residents to adopt a voluntary quarantine, then instituted a mandatory one, closing churches, schools, and theaters with little warning. But they could do only so much. By the end of February 1919, more than 1,400 would die in Seattle alone.

When Thomas Riggs, governor of the Alaska Territory, heard the disease had reached the West Coast, he was desperate to prevent its spread to the north. He implored steamer companies headed to Alaska to inspect their passengers for signs of the flu. Physicians were assigned to Alaska ports, ready to examine anyone who appeared sick and, if necessary, to put them in quarantine. On October 14, 1918, the first case was reported in Juneau, on Alaska's southeastern panhandle, and the illness quickly spread along the coast. Still, southeast Alaska came through the epidemic relatively well: Doctors set up makeshift hospitals, the dead were buried immediately, and aid came into the region from as far away as San Francisco, all of which helped to slow the spread of the disease. The same could not be said for the northwest region of the state.

The *Victoria* steamed into Nome, about a hundred miles southeast of Wales, on October 20. Much of what happened next is known because of an exceptional series of stories from the *Anchorage Daily News,* which chronicled the spread of the illness in northwest Alaska. Everyone on board the ship had been examined in Seattle, and none had shown symptoms of the disease. Still, when they arrived, a doctor ordered them quarantined at the local hospital. After five days, just one person had taken ill. Doctors dismissed it as a case of tonsillitis, and the quarantine was lifted. Four days later, a worker from the hospital got sick and died. Two days after that, all of Nome was placed under quarantine, but it was too late.

Some estimates say more than half of the white residents of Nome were sickened. More than two dozen of them had died by the last week of November. It was even worse in Nome's Native village, slightly removed from the town, where 162 people died in an eight-day period. Panicked, many of them had run from house to house, intending to warn others but in the process spreading the disease to nearly every household. Later, when rescuers went to check on cabins whose chimneys showed no smoke, they found whole families frozen, unable to summon the energy to keep a fire going. Several Natives, amid all the destitution, took their own lives. The town's chief physician, Daniel S. Neuman, fell ill within two weeks of the first case and was incapacitated, leaving just one physician at nearby Fort Davis to tend to the scores of sick. The ship that had brought the flu was the last of the season. There would be no one coming to help them.

When the *Victoria* put in at Nome and its passengers were quarantined, the mail unloaded from the ship was fumigated, but the crew came into contact with local mail carriers before they loaded their dogsleds with letters and packages bound for the isolated communities on the coast. In early November, a mail carrier and his son were mushing toward Wales when the boy got sick. Arthur Nagozruk Sr., head teacher at Wales and the de facto community leader on the cape, had heard about the flu and given the mail carrier strict instructions: If either he or his son was sick, they were not to return to the village. Maybe he thought someone there could help his son, or maybe he was too despondent to turn back, but the mail carrier ignored Nagozruk's command. By the time he rode into Wales, his son was already dead. Two days later, the father fell ill. Within a week, nearly the entire village was sick.

A government nurse tried to help, but she didn't have enough medicine to treat everyone, and her supply of food was quickly exhausted. She ordered the boys to slaughter reindeer, feeding the meat to the children and making broth for the babies. People flocked to the school, and many of them died there, their bodies stacked in neat rows in a back room. Others never had the strength to leave their homes and died in their sod houses. When rescuers arrived a few weeks later, they found babies suckling at their dead mothers' teats and children using their own body heat to thaw tins of milk to feed even younger children.

Before it was over, 170 people—roughly half of the village— would die. Among the dead in Wales were almost the entire village council, two Native teachers, and most of the whaling crews. Five babies who were born around the time the epidemic started died before it was over. Nagozruk, the leader of the village, lost his wife and two sons. Two of Gene Agnaboogok's grandparents were among the victims, thirty-nine years before Gene was born.

When a rescue crew arrived in Wales in the spring of 1919, they dynamited holes just up from the beach north of the village and buried the dead en masse, interring them in the dunes under a white cross where the cemetery still sits to this day.

A year and a half later, Henry Greist, a Presbyterian missionary, came to Wales, where he lived for a year before going to build a hospital on Alaska's northern shore. In an unpublished manuscript, he described what happened next, after the rescue mission arrived on the cape. He got a few things wrong—namely the year, population, and death toll of the flu—and he harbored some abhorrently racist views toward the Natives, but his is the only known written account of what he described as a "tragical sociological experiment."

One of the folks in the rescue party was an official from the U.S. Bureau of Education, likely Ebenezer Evans, who went to Wales in early 1919. He knew that there were as many as forty orphans left in the village, and he came armed with a sheaf of marriage licenses. Gathering the widows, widowers, and other adults of marriageable age in the schoolhouse, Evans gave them a choice. The orphans would be shipped out to towns and communities untouched by the flu, lost forever not just to the village but to their remaining relatives as well—aunts and uncles, brothers and sisters. The only alternative, Evans told the assembled crowd, was for them to "here and now" choose a new spouse and be married on the spot.

The survivors—many of them still grieving for loved ones who lay dead in the next room or who were, at that moment, being buried outside by strangers in a mass grave—were, by Greist's account, horrified at the thought of losing an entire generation of children after the great loss they'd already suffered. At Evans's urging, they lined the walls, men and boys on one side, women and girls on the other. Each man was then told to pick a wife. If they hesitated, one was chosen for them. One older man, a wealthy fifty-year-old, stood with downcast eyes. His wife, an intelligent and capable woman whom he had loved deeply, had been taken by the flu, and he was still grief-stricken. When he refused to pick, Evans paired him with a girl only a third his age. Not only was the girl much younger than him, but she was already in a relationship with a boy her age who wasn't at the schoolhouse that day. She sobbed uncontrollably as she was led to a table to sign a marriage certificate. Once every man and woman were paired off, Evans presided over a mass ceremony, pronouncing each couple man and wife.

Soon after the ceremony, the older man showed up at the door of his new wife's mother. He left the young girl at the doorstep, later obtaining a legal divorce. The girl would go on to marry the young man she'd intended to be with. The older man remained unmarried for the rest of his days.

The flu wreaked havoc among the Native population of Alaska. Of all the people who died in the territory in 1918 and 1919, half died of influenza, more than 1,100 in total. More than 80 percent of them were Native. Two-thirds of the dead came from the area in and around Nome, including the village on the cape.

Wales would never be the same. The population had been reduced by almost half and its entire familial structure involuntarily rearranged. Once larger than any other Native community on the coast, Wales would never again see a population of more than 160 people. But the loss for the village went far beyond the body count or the changing of last names on a piece of government paper. Reindeer herders and leaders and the village's most capable hunters had died. The dead took with them the oral histories that had been passed from elders to young in the warmth of the *qargis,* the knowledge of how the currents flow in the strait, of when the ice comes, where the walrus congregate, and how to know when a storm is approaching. Their resilience had come from their stories, and in an environment as dangerous as the far north, the knowledge embedded in those stories could be the difference between life and death. Whaling would be all but abandoned for half a century, and the loss of so many hunters left Wales residents more reliant on food shipped in from Nome and beyond.

The Natives of Alaska did not live in a utopia before white people arrived. There were hardships and conflict and illness and hunger, just like in any community. But the changes brought on by the arrival of the outsiders was unlike anything they had seen before. The process of stripping the Natives on the cape of their tradition and culture began when the first missionaries landed on the beach in 1890. Where the Natives saw the world around them as something with which to work in concert, the white men saw every natural resource as an opportunity, a herd to be imported, a whale to be harvested for oil, a people to be converted into Christian workers.

It wouldn't be the last time their way of life was threatened by forces beyond their control.

Chapter 7

Milestones

On the internet, Nora was a star. After seeing the video of Cindy Cupps, the veteran keeper, feeding Nora as a newborn, thousands tuned in to Facebook and YouTube to watch the cub's development. But her fans were unaware that the keepers harbored a nagging concern. Despite everything—the extensive research, the careful monitoring, the round-the-clock care, the twenty-eight-hour shifts, the missed holidays—the Nora Moms had fallen short.

In the zoo hospital, where they'd learned that Nora wasn't absorbing enough vitamin D and calcium and that metabolic bone disease had left parts of her skeleton malformed, veterinarian Priya Bapodra had dried her tears and pored over the research to figure out how to balance the cub's diet. She suspected the problem had begun when the team switched the fat in Nora's formula from safflower to herring oil. The research had suggested herring oil would be better, but that hadn't turned out to be the case for Nora.

Nora's keepers switched her back to safflower oil and added extra calcium to her formula. She also started getting vitamin D injections. Nora wasn't in obvious pain, but polar bears don't always show when they're hurting, so they gave her light painkillers, too. When the Moms took her out of the incubator, they tried not to handle her too much, so her bones could heal.

The keepers would never know why Aurora abandoned her cub, why Nora didn't get the nourishment she needed directly from her mother. Maybe Aurora sensed that the baby bear was fragile. Perhaps she saw the death of Nora's twin, just a day after his birth, and instinctually knew that caring for a sickly cub could put her own survival at risk, as it would outside the confines of captivity.

In the wild, Nora would have snuggled close to her mom in the den as she adjusted to the cold of the Arctic. Over the course of several weeks, the Moms gradually lowered the temperature in the ICU from eighty-eight to seventy to fifty-five degrees. They brought electric blankets, beanies, and thick sweatshirts to stay warm. They rested on an air mattress in the corner so they could be close to Nora all the time. They dressed in thick Carhartt outfits to protect themselves from her curved claws and new teeth. She would appear content one second, and then an unfamiliar noise would set her off. Under different circumstances, her mother might have set her straight with a firm paw.

Nora still pulled herself along like a seal, back legs splayed behind her. She'd push herself up and then topple and roll, no particular structure to her, just fur and feet and dark, curious eyes. While the Moms worked on email or entered data in the corner of the darkened ICU, Nora would scoot over to them and wrap her paws around their ankles. They were always strategizing, always worrying, but there were moments when they stopped and just stared at her, lovestruck. Shannon Morarity sometimes held Nora while the cub napped, and at least once she nodded off herself, snuggling a warm armful of polar bear. In a holiday-themed video, released just after Christmas, Morarity fed Nora from a bottle, gingerly holding the bear's injured forearm, before the cub seemed to involuntarily doze off.

Babies recover quickly, and it took only a few weeks for some of Nora's bone issues to right themselves. Bapodra and the rest of the Moms were overjoyed at how fast she seemed to improve. They gave her one more vitamin D shot on January 21. X-rays taken two days later showed the fracture in her forearm had healed. Nora appeared to be a healthy baby bear.

At the end of January, Nora lapped water from a green bowl, and the Moms cheered. A couple of days later, they gave her a tub of water to explore, and she parked her rump in it.

The head veterinarian told Bapodra what she already knew: "You're going to have to let her go."

Nora was medically cleared, and it was time to move her from the animal hospital to Polar Frontier, where the other bears lived.

The Polar Frontier exhibit at the zoo in Columbus is modeled after an abandoned Arctic mining town, reminiscent of what Nome might have looked like after the gold rush swept through in the early twentieth century. On the north side of the faux-weathered wood-planked main building is a pen where Arctic foxes scamper back and forth. Beyond that, the grizzly enclosure, where brothers Brutus and Buckeye share a third of an acre. Next door to the grizzlies, the polar bear enclosure is modeled to look like Arctic tundra, with short grasses, rocks, and sandy areas surrounding a massive swimming pool, which is stocked with live trout and heated and cooled by geothermal energy. The viewing space features a wall of windows more than a hundred feet long and stairs down to a subterranean level, fashioned to look like a mine shaft. At the bottom of the pool, clear plastic lets zoo-goers watch the bears swim and dive from directly underneath—a sort of glass-bottomed boat in reverse. It's not hard to look up at the silhouette of a bear preparing to jump and imagine it's the last thing an unlucky seal might see in the wild.

The exhibit had gotten a $20 million facelift in 2010, and it could host multiple bears in separate areas. The Columbus Zoo was now home to three adult polar bears, plus Nora—more than any other zoo in the country at the time—and at 1.3 acres, the enclosure was spacious enough for all of them, a reflection of the zoo's commitment to the species.

Visitors to the exhibit saw few walls, in keeping with modern zoo design. It appeared as though Nora could wander north, leave the zoo, hang a left when she got into Canada, and join her Alaskan kin. She explored her new surroundings tentatively, looking back to Morarity and the rest of the keepers every time she encountered something new.

After Nora had spent about two weeks in Polar Frontier, Morarity and another keeper pulled on wet suits and waded into the seven-foot-deep pool to show her it was safe. The cub stalked the edge of the water before dipping in a paw. A few minutes later, she waded in. She was a natural swimmer. Seeing her take off in the water, paddling with her outsize paws, assured them that there were things about being a bear that, somehow, Nora already knew.

In the log the Moms kept of Nora's milestones, the "swimming" entry is accompanied by four exclamation marks.

As Morarity watched Nora in the yard, she felt her shoulders relax and thought, *Okay. She's good.* Getting her to this point had been exhausting. Now the bear had room to play with all five of her Moms at once, and they romped with her in the yard like a bunch of kids.

There was a risk to the bonds the keepers had built with Nora, though, and they knew it. Birds and mammals are hardwired to gain their sense of self early in their cognitive development. In the 1930s, Austrian scientist Konrad Lorenz theorized that baby geese would seek out and identify with the first moving object they saw after they hatched, whether it was another species of bird or a stick or Lorenz himself. He was right, and ended up on the covers of magazines with a flock of baby birds following in his footsteps. "The man who walked with geese" won a Nobel Prize in physiology for his work. Imprinting, as the process became known, was valuable for those working with domestic fowl. The birds could be imprinted on inanimate objects, like colored balls or electric trains, and be led about and trained to behave.

But there were drawbacks, too, especially for wild animals. A wild bird imprinted on anything other than its mother would never be able to function properly with other members of its species. Some migratory birds won't migrate, and birds of prey won't learn to hunt. If you find an orphaned bird in the wild, it's typically illegal to raise it yourself, in part because of the risks imprinting poses. Biologists who raise birds for reintroduction to the wild take great pains to avoid it, donning sock puppets that resemble whatever species they're working with. At the Monterey Bay Aquarium, in California, workers wear Darth Vader–like costumes to keep baby otters from imprinting on them, and the internet is full of whimsical photos of zoo workers dressed in full-body panda suits cradling endangered cubs. We know a little more about imprinting today than Lorenz did back in the thirties, but the risks remain the same, and they worried the Nora Moms. Nora could sense the other polar bears at the zoo, through sight and sound and her powerful nose, but her primary exposure was to people. There wasn't much the keepers could do about it, though. Nora was still too small to be in the same enclosure as the other bears, and Aurora had already shown that she wouldn't protect the cub from danger. The young bear had a lot to learn, and only humans to learn it from.

When Nora reached five months, the keepers had to pull back. Nora's claws and teeth had gotten longer and sharper. She was too big and too strong to be with humans, but too small to be with the other bears, who could easily injure or kill her. The Moms interacted with her through glass or netting. They still talked to her and told her they loved her. But inside the enclosure, she had only the sounds and smells of other bears to keep her company.

When Nora roamed the yard, Morarity stood at a door where the cub could see her. For a while, Nora looked to Morarity every time a new noise or a strange smell wafted across the enclosure. And then one day she stopped looking. It was the moment all moms work toward, and all moms dread. The keepers would never get to cuddle her nose-to-nose again.

But soon Nora would have a whole world full of people lined up to see her from the other side of the glass.

Animals have been put on display for human entertainment for thousands of years. Rulers in ancient Egypt kept collections of elephants, baboons, and hippopotamuses as early as 3500 B.C.E. In the eleventh century B.C.E., King Ashur-bel-kala of the Middle Assyrian Empire built massive botanical and zoological gardens to complement his cadre of creatures. Roman emperors kept hundreds of animals, too, some used to kill criminals in public executions held in front of tens of thousands of people. Early menageries weren't for public consumption. Rather, the animals were kept behind closed doors, to be viewed by, and traded among, the rich and powerful. More than displays of wealth, the animals represented dominion over the natural world. The more exotic the creature, the greater the dominion. Polar bears, hailing from a nearly inaccessible corner of the globe, were among the most sought after.

King Henry III kept one, muzzled and chained, in the Tower of London in the thirteenth century. Five hundred years later, Frederick I, the first king of Prussia, is thought to have kept a polar bear among his sprawling collection of exotic creatures, which he would pit against one another in battles. The animals were hard to come by, and too valuable to watch die, so he had their claws and fangs removed before they were dumped into their fighting enclosures.

Soon after Frederick's foray into captive animal fights, the traveling menagerie was born. Animals were loaded into cages and trucked about the countryside in carts. The public paid to view the creatures from faraway lands, displayed prominently at carnivals and fairs.

In the early 1800s, the Zoological Society of London began building one of the first recognizably modern zoos in the West. Located in the city's Regent's Park, the London Zoo was originally created as a place for scientific study, closed to the public. By the middle of the century, though, the public was allowed in, and over the decades the zoo was redesigned to cater to London's large population, with wide pathways and ways to easily view the animals.

For more than a century after the founding of the London Zoo, zoos existed as places to observe animals and analyze their behavior, but their owners weren't overly concerned with conservation or with the well-being of the creatures they kept. Often, the animals were trained to perform for audiences, usually by abusing them into docility. That began to change in the United

States in the 1970s as policy makers and activists pushed for a philosophical shift, culminating in the passage of the Animal Welfare Act. Today, zoos are billed as places not just of science but also of education and environmental awareness.

For most of us, zoos represent our only chance to see first-hand the splendor and diversity of the natural world. Without zoos, most children would never witness the graceful gait of a tiger or appreciate the dexterity of an elephant's trunk. Few people will work to save something they've never seen, and were it not for the viewing platforms at primate exhibits in zoos around the world, only the privileged few would ever be able to look into the eyes of an orangutan and see something undeniably familiar staring back. As one menagerie owner put it around the turn of the eighteenth century, he was doing "more to familiarize the minds of the masses of our people with the denizens of the forest than all the books of natural history ever printed."

That's still essentially the argument for zoos. As more species have come under threat, from hunting or loss of habitat or environmental changes, zoos have become more than just a place to learn about animals. As their public relations staffs will tell you, they've become an important tool for preserving species on the brink of extinction.

But no animal's ideal home is in captivity, especially a polar bear's. Getting the climate right is tough. Zoos build enclosures with fans and give the animals indoor areas with air-conditioning. They keep their pools cool and ply them with fish frozen into blocks of ice. Even in their native habitats, some polar bears can see summertime temperatures well into the eighties, so they are built to endure, or at least tolerate, brief stints of warm weather. But the weather in central Ohio rarely resembles that of the Arctic.

Diet presents its own problems. Zoos are marketed primarily

as family-friendly destinations, and few parents want to explain to their young children why the cuddly polar bear just ripped apart a seal right in front of them. That leaves people like Dana Hatcher, the nutritionist who came up with Nora's formula, to try to replicate their wild diets in other ways. Zoos also have to engage the minds of the animals they care for. A bored animal is an unhappy animal, and keepers go to great lengths to provide enrichment. Nora's keepers created training routines and puzzles she had to solve for rewards. They encouraged her to play, and they moved things around in her habitat to keep her from developing a routine.

American zoos are regulated by the U.S. Department of Agriculture, which provides licenses to any facility aiming to display animals for entertainment and is charged with enforcing the Animal Welfare Act. But even licensed zoos run the gamut, from facilities like the one in Columbus, which offers its polar bears an expansive, multi-million-dollar habitat, to others that provide little more than a lukewarm pool and an iceberg mural. The real mark of a respectable facility is accreditation from the Association of Zoos and Aquariums, which holds zoos to a higher standard of care than the law requires. Of the 2,800 licensed animal exhibitors in the United States, fewer than 10 percent are accredited. The ones that are wear it like a badge of honor, with plaques at their gates and AZA logos on their websites.

Management of endangered species gets extra consideration within the association. Each such species has its own Species Survival Plan, written by elected members of committees from member zoos. The committees make recommendations for where the animals should be placed and when keepers should encourage breeding. For polar bears, of which there are relatively few in captivity, protecting genetic diversity is especially critical. After Gene Agnaboogok accidentally orphaned him in Alaska, it was those types of considerations that dictated Nanuq's moves around the country, from a zoo in Anchorage to Wisconsin to New York to the Polar Frontier exhibit in Columbus, where he fathered Nora in late 2015.

On a warm day in mid-April, a line of eager zoo-goers stretched through Polar Frontier, past walls plastered with facts about climate change, past a time-lapse projection showing shrinking sea ice, past posters urging people to conserve fossil fuels, and ending at the floor-to-ceiling glass that looked out onto the enclosure, where throngs of adults with fancy cameras and children with expectant smiles waited for Nora.

She strutted into the yard, seventy pounds of pigeon-toed fluff, and immediately got her head stuck in an orange traffic cone, tromping around like a confused construction worker in a fur coat. She played with a yellow ball. She buried her nose in a pile of ice chips. She belly-flopped into the pool, scattering fish.

"Do it again!" someone shouted. "Jump again!"

Nora's debut was covered by nearly all the local media. Shannon Morarity, the zookeeper who had been with Nora for all but a few of the cub's first 159 days, teared up on Channel 10.

"We're proud of her," she said.

Over the next five months, the zoo counted 261,126 people who came through the Polar Frontier line to see Nora. Some ran through the gates. Some visited so often that the keepers recognized them. Nora, for her part, appeared to thrive on the attention. When she was on exhibit, usually for just a couple of hours a day, she spent much of her time in front of the window that separated her from her fans. She seemed to love people, and they loved her back.

Nora's keepers had scoured the research on how to care for

her, and the cub seemed happy to them, if the happiness of another species is a thing that can be judged. She would never have to roam the vanishing sea ice in search of food or be forced to rummage through garbage or raid goose nests or eat berries in lean times, like her northern cousins. The keepers known as the Nora Moms would exhaust themselves to attend to her needs.

But could they offer her enough open space? Were traffic cones and oversize toys enough to engage her mind? What about her social needs? In the wild, young bears are completely reliant on their mothers. Mature polar bears spend months alone as they search for food, but when their needs are met, they have been known to congregate with other bears. In her new enclosure, she would be exposed to others of her species and the grizzlies in the neighboring yard—able to see and hear and smell them—but Nora had never actually been in the physical presence of another bear since Aurora left the den months before, before her eyes were even open.

After five months with the keepers, it was unclear whether she knew to which species she belonged.

Chapter 8

Farewell

Aside from the bears she could see and smell through the doors of her enclosure, Nora was alone. Except, of course, for her legions of devoted fans. Fans from all over Ohio who came to see her regularly. Fans from across the country who got on trains or buses or airplanes to watch her play. Fans around the world who waited for new videos to appear on YouTube, who flooded the Columbus Zoo's Facebook page with adoring comments. Hundreds lined the glass wall on the west side of the big yard where she played. Thousands poured into the zoo every week, standing in line to get into the Polar Frontier exhibit. Millions more followed her online.

Nora was beloved.

She was far from the first superstar in the zoo world, though. There was Heidi the cross-eyed opossum, who won over the local press and then the global internet in 2010 after she was photographed at the Leipzig Zoo, in Germany. There was Koko, a western lowland gorilla born at the San Francisco Zoo in 1971, who was said to have a thousand-word vocabulary in what her keepers called Gorilla Sign Language. The 280-pound primate adopted and cared for kittens, which she gave names like All Ball, Lipstick, and Smoky. Harriet, a Galápagos tortoise who was about five years old when she was allegedly taken from the islands by Charles Darwin, lived to be an estimated 176 years old before her death in 2006, her fame sealed by having existed for nearly two centuries.

And then there was Knut.

Born in December of 2006 to Tosca, a twenty-year-old mother rescued from a circus in East Germany, Knut was the first polar bear born at the Berlin Zoo to survive past infancy in three decades. Like Nora, Knut had been abandoned by his mom just a few days after he was born. Scooped out of the enclosure with an extended fishing net, Knut became a media darling almost immediately. He was raised by a keeper named Thomas Dörflein, who fed, bathed, slept next to, played lullabies on his guitar for, and appeared with the cub on an almost daily basis. Dörflein became a bit of a celebrity in his own right, earning Berlin's Medal of Merit for his care of Knut. Within a few months, though, the baby bear had gathered almost as much controversy as he did attention. Animal rights activists said the zoo should have let Knut die instead of trying to raise him like a small furry person. Wolfram Graf-Rudolf, director of a nearby zoo, agreed, telling a German newspaper that Knut's keepers "should have had the courage to let the bear die." The risk, argued Graf-Rudolf, who had seen animals hand-reared at his own zoo, was that the cub would become too attached to Dörflein and develop neurosis. "Each time his keeper leaves him, and he can't follow, he will die a little," he told BBC News. The group People for the Ethical Treatment of Animals sued the zoo for "extreme animal mistreatment."

Berliners were having none of it. Some of the earliest coats of arms for the city, dating back to the 1200s, depict a bear, and the German capital still uses one as its emblem. The people of Berlin loved their little cub, and they flooded the zoo with emails and letters pleading for Knut's life as children protested at the zoo gates. The zoo obliged, sanctioning Dörflein's care for the cub and subsequently turning Knut into an international brand all his own. His public debut was covered by no fewer than four hundred journalists from as far away as Uzbekistan and South America. He had his own blog, written in the first person and translated into several languages, as well as a webcam that streamed video from his enclosure. Attendance soared after Knut's debut, setting records at the 163-year-old zoo as Dörflein and the cub made twice-daily appearances for the throngs of fans that packed the exhibit. Annie Leibovitz, the world-renowned photographer, took portraits of the bear for *Vanity Fair,* and Knut appeared on the cover in May 2007, edited into a picture of Leonardo DiCaprio standing on an iceberg for the magazine's Green Issue. Every aspect of the cub's life was under a microscope. Wild speculation about his health erupted after he was taken out of the exhibit for teething pain. One visitor claimed to have a blurry picture capturing the moment Knut was conceived.

The hubbub was too much for some, and "Kill Knut" graffiti started appearing around the city. Police were called in to protect the bear after an anonymous death threat was faxed to the zoo.

But even as some soured on the bear—"Knut Steadily Getting Less Cute," read a *Der Spiegel* article in April, when he was just five months old—his marketability continued to grow. The zoo registered the cub's name as a trademark, and shares in the facility doubled in value on the Berlin Stock Exchange. There were Knut ringtones and Knut coins and a Knut TV show. The government got in on the action, too, issuing a Knut postage stamp. The zoo, which was forced to limit visitors to seven minutes of viewing time, pledged that most of its proceeds would go toward conservation. As Knut aged and lost some of his infant qualities, the crowds dwindled, but he kept a robust fan base both locally and internationally, and the zoo continued to see record numbers of visitors. The zoo that owned Knut's father sued the Berlin Zoo for proceeds stemming from the cub's fame. A protracted legal battle played out and the Berlin Zoo eventually paid nearly $600,000 to keep the cub in the German capital.

By July, Knut weighed more than a hundred pounds, and the zoo curtailed his contact with people, out of safety concerns for his keepers. He needed his independence, too, a zoo spokeswoman said, adding that it was time for Knut to "associate with other bears and not with other people."

Markus Röbke, one of the keepers who helped raise the bear, said it was clear that whenever Dörflein was away, Knut missed his father figure, and that the cub was highly dependent on the attention of his adoring fans. He howled whenever he caught Dörflein's scent and cried if the windows in front of his enclosure were empty. When the zoo had to close for a day because of icy conditions, he whimpered for hours until a keeper, out of pity, stood in front of the glass to quiet him. The zoo banned employees from playing with the cub, and Röbke argued that Knut needed a change of scenery to properly develop. "He doesn't know that he's a polar bear. As long as he's with us, he will always regard Thomas Dörflein as his father," he told *Der Spiegel*. "Knut needs an audience. That has to change."

In September of 2008, Dörflein died of a heart attack, and eventually Knut was moved to an enclosure with three adult female bears, including his mother, Tosca, but he was never fully accepted. He spent most of his time alone, and news reports said the other bears acted aggressively toward the cub, even calling the behavior "bullying" after video emerged of one of the females lunging at Knut and pushing him into the water. Roughhousing is not uncommon in polar bears, and his keepers downplayed the incident. "For the time being, Knut is not yet an adult male and doesn't yet know how to get respect like his father did," one keeper told *Time* magazine. "But day by day, he is imposing himself and with time, this type of problem will go away."

He would never get that chance.

On March 19, 2011, Knut, who had been standing on a rock next to one of the pools in his enclosure, started walking in circles. He spun several times before his back leg began to shake. Convulsions rippled through his body and he collapsed backwards into the pool as the assembled zoo visitors cried out. Knut drowned in the pool in front of hundreds of his fans. An autopsy later revealed that he was suffering from encephalitis and probably had been for weeks before his death; the infection, which causes swelling in the brain, likely would have killed him even if he hadn't drowned. Though he was years removed from his adorable cub stage, the outpouring of grief over Knut's death nearly rivaled the joy that had surrounded his birth. Flowers piled up near his enclosure as hundreds visited to pay their last respects. "He had a special place in all of our hearts," Berlin mayor Klaus Wowereit told a local paper. "He was the star of Berlin Zoo." Today, a life-size bronze statue of Knut straddles two faux icebergs on the zoo grounds. Knut himself was taxidermied, his stuffed body added to the collection at Berlin's Natural History Museum. Even his final resting place raised the hackles of his fans, though. Thousands signed an online memorial book, many demanding that the bear receive a proper burial, some calling for the resignation of the zoo director.

For all their differences, Nora and Knut share indisputable commonalities. Raised by people, both bears grew up under a spotlight thrust upon them because of the hardships they had endured. They were cute, of course, but both of them also offered a story of overcoming rare and unique challenges. Those stories were good for the zoos, too. The miracle bears that survived being abandoned by their mothers would get people to buy tickets, but they could also become ambassadors for their species. Gerald Uhlich, a trustee on the Berlin Zoo's board, said, "Knut has become a medium of communication. . . . [He] will be able to draw attention to the environment in a nice way. Not in a threatening, scolding way." The German bear was adopted as the official mascot of an endangered species conference, and on the stamp bearing his likeness, his face was framed with the words PRESERVE NATURE WORLDWIDE.

Nora played that role, too. Visitors to Polar Frontier walked over interactive displays on the floor, with projectors displaying ice that cracked when they walked on it. They walked past signs that reminded them they could shrink their carbon footprint by turning down their thermostats. They listened to docents who told them about shrinking sea ice, in between tidbits about Nora's diet and upbringing. When she was just over eight months old, about four months after she went on public display, Nora got a visit from the State Department's special representative for the Arctic, Robert Papp. "There's the abstract and the theory and the education, but then there's the reality of the experience," he said after visiting with Ohio's favorite bear. The zoo was bringing the far-flung realities of climate change home to the people of the Midwest, he said. "Here you have a chance to get a sense of it, to feel it a little bit. The more we can expose people to this reality and the challenges the animals face out in the wild and the reality of the impact human behavior and existence is having on that environment, we gain believers. We gain people that perhaps in the future will work for change and to help us deal with this monumental problem that we're facing on this planet."

Presenting the public with an amorphous problem, with ambiguous impacts that are hard to see and comprehend, isn't exactly inspirational. Show them a baby polar bear who has already overcome tremendous adversity, and remind them that bears like her are suffering in the wild, and perhaps they'll be inspired to act. The long lines of paying customers didn't hurt—most zoos will be happy to tell you how much of their proceeds goes toward conservation projects—but connection with the animals was the main point. During the course of her roughly six months in front of the Ohio public, Nora forged that connection with more than a quarter million people who visited her at the Polar Frontier exhibit.

But it wasn't enough.

Far to the north, on the real polar frontier, the Arctic was changing. In March of 2016, northern sea ice reached its maximum extent, the point at which as much of the region that's going to freeze that year has frozen. On March 24, experts measured the ice extent at more than 5.6 million square miles, stretching from well south of Wales, on Alaska's west coast, up through the Chukchi Sea, covering the entirety of the Arctic Ocean, and extending icy fingers that wrapped around the islands of the Canadian Arctic Archipelago. It was a lot of ice by any measure, except one. Up to that point, it was the lowest maximum extent of Arctic sea ice researchers had seen since satellite records began, 7 percent less than the thirty-year average. It was five thousand square miles less than the previous record low, which had occurred just the year before.

Arctic temperatures had been above average from December through February, and in March things got worse. Winds out of the south pushed the ice edge off Russia's north coast, while warm-water incursions from the Atlantic came in through the Norwegian Sea, eating away at the ice from above and below. The first two weeks of March saw temperatures at the North Pole eleven degrees Fahrenheit above normal.

A cascade of dire news about the impacts of climate change had been tumbling forth. Just a few weeks before Nora met the public, the Obama administration had released a three-hundred-page report on the health effects of climate change. Allergy season would likely set in earlier, as would the season for Lyme disease as warmer temperatures allowed ticks to expand their range. Thousands of Americans, possibly tens of thousands, would die prematurely every summer as heat waves increased in intensity, frequency, and duration. Increased runoff from extreme rains would warm bodies of water, making them more susceptible to toxic algal blooms, bolstered by fertilizer washed downriver by massive storms. Everyone would be affected, the report said, but not equally. Women, the poor, Indigenous communities, minorities, immigrants, and the elderly would bear the brunt of what was coming. Those who had done the most to create the problem—the executives of companies that had continued to exploit the world's resources even as they knew the havoc they were creating—would be able to buy themselves some safety, at least in the near term.

But public awareness of environmental issues had come a long way since James Hansen testified before Congress on that hot day in 1988.

The Intergovernmental Panel on Climate Change (IPCC), the United Nations body formed in the wake of the Hansen testimony, had released its first report in 1990, with successive reports coming every five or six years. The panel's charge was not to do original research, but to synthesize what was known about climate science to inform the decisions made by those in power.

The first report largely echoed what Hansen had told Congress, but with more confidence. It set the stage for an international agreement at the United Nations Conference on Environment and Development, better known as the Earth Summit, in Rio de Janeiro in 1992, with each of the signing countries pledging to reduce its greenhouse gas emissions. The agreement also set forth a plan to meet on a yearly basis to assess progress. It was all nonbinding—no country really *had* to do anything—but with 177 world leaders signing on to the pledge, it was the widest acknowledgment to date of the scale of the problem the world was facing.

There had been pushback on climate science since the days of Jean-Pierre Perraudin and Louis Agassiz, mostly in the form of healthy scientific skepticism. But the late 1980s saw a different kind of resistance begin to take shape. The Global Climate Coalition (GCC), a nonprofit representing more than forty corporations and trade associations—including Shell, Chevron, Ford, Chrysler, and the American Petroleum Institute—was one of the earliest to attempt to undermine the findings of climate researchers. With an army of lobbyists and public relations experts, coffers full of fossil fuel money, and a strategy borrowed from the defenders of Big Tobacco, the GCC worked to get its own scientists installed on IPCC panels to downplay the threat of climate change. Its lobbyists met with UN scientists and tried to get industry language inserted into official reports. The coalition wrote that "the role of greenhouse gases in climate change is not well understood," even as its own researchers wrote the exact opposite in internal reports, saying the science of climate change was "well established and cannot be denied."

The campaign against substantial regulations of carbon didn't aim to completely win anyone over. It didn't have to. Groups like the GCC only had to sow enough doubt to keep people from taking meaningful action. For the most part, the gambit worked.

After the Earth Summit agreement, the international community, including President Bill Clinton, signed on to the Kyoto Protocol in 1997, but the Republican-held Senate made sure it never came up for a vote and Clinton's successor, President George W. Bush, rejected it, at least in part because of lobbying from the GCC. Meanwhile, the UN panel released report after report, each one painting a dire picture of the future if carbon emissions continued unchecked. The GCC's influence would soon fade as many of its biggest members withdrew from the group, but the damage was ongoing, and any gap they left in the climate denial space was quickly filled by other well-funded groups.

By the time the UN released its fourth climate report, in 2007, it was clear that not all parts of the earth were warming at the same rate. Over the previous century, the Arctic had warmed at roughly twice the rate of the rest of the planet, the report said, with some projections predicting that summers in the far north would be nearly ice-free by the end of the twenty-first century. In the fall of 2012, the Arctic saw its lowest minimum sea ice extent on record. The minimum extent is arguably more important than maximum extent, because ice that survives the summer becomes multi-year ice—less salty and therefore stronger than first-year ice. But the operative word is "survive," and 2012's minimum extent was 293,000 square miles less than the previous record low. It was roughly 1.27 million square miles below the average from 1979 to 2000. The Arctic was missing enough ice to cover the entire nation of India.

Over the summer of 2016, Nora continued to tick off milestones. About two weeks after she met the public, she got her whole body underwater for the first time. A little more than a week later, she was regularly diving three or four feet down and caught her first prey, one of the darting fish the zoo stocked her pool with. By the end of the summer she was diving all the way to the bottom of the pool, a distance of twenty feet. She was increasingly comfortable seeing the other bears from a distance, and even meeting them through screens in the heavy steel doors, out of view of the public.

But Nora would never be reunited with her mother, Aurora. As much as Nora needed the companionship of other bears, and as much as the zoo adored her, it didn't make sense for her to stay. In the spring, the keepers had seen Nora's father, Nanuq, mating with Aurora and the zoo's other female bear. If they were pregnant, they'd be giving birth in November or December. The zoo had to make room, and Nora was occupying the space they needed.

The Association of Zoos and Aquariums, the organization that recommends placement for bears, and the committees of the Species Survival Plan within it looked at every zoo in the country capable of housing a cub. They needed a zoo that not only had enough space for the growing cub but could also offer the companionship of another bear. Nora seemed the perfect company for Tasul, an older bear that lived at the Oregon Zoo, in Portland.

Late in the summer, Nicole Nicassio-Hiskey, one of Tasul's keepers, traveled from Portland to Columbus to meet Nora and her Moms. It was Labor Day, Nora's last day in front of the public. The Nora Moms all came out, decked in matching pajamas covered in polar bears. There was a festive vibe, like at a going-away party for a freshman headed off to college. On one of the rocks in the yard, one of the keepers had used chalk to scrawl

NORA: ALWAYS A BUCKEYE, a reference to the mascot of Ohio State University, located in Columbus. Nicassio-Hiskey helped stock the yard with toys while others decorated with streamers and signs reading BON VOYAGE! By then, many of the bear's fans felt as if they knew her. They'd watched her grow. Nearly seven thousand of them showed up on her last day. As Nicassio-Hiskey walked through the exhibit, a group of volunteers gathered around her, eyes fixed on the Oregon Zoo logo on her shirt. They peppered her with accusatory questions, some jokingly, some not so much. "So you're the one taking our Nora?"

For the Moms, who'd given much of their lives to Nora for the better part of a year—birthdays and holidays and marathon shifts in the intensive care unit—that they always knew she'd eventually be leaving didn't make it any easier.

Just a few days after the public bid Nora farewell, Shannon Morarity stood on the runway of the Indianapolis International Airport, in front of a FedEx cargo plane. Nora sniffed the unfamiliar smells of the airport from inside her shipping crate next to the hangar. She'd been weighed and checked in and given a mild sedative. Morarity and another keeper took turns saying goodbye, feeling equal parts sad and optimistic. They'd done the nearly impossible. This ten-month-old bear, who faced long odds of survival, was starting a new life.

"We love you," Morarity told Nora through the metal bars. "You have so much to look forward to, and we're so excited for you."

The other keeper stood by, taking a video with her phone. They were proud of the job they'd done, the five women who had looked after Nora since the day she was born. The whole zoo had come together to cover their shifts while Nora pulled them away. If so many people could devote themselves to just one long-shot baby bear, maybe the planet could pull together to save the rest of her species. A curator and a veterinarian hopped on the plane to accompany the young bear for her flight. Morarity watched the door close. As the plane lifted into the sky, Nora let out a low, rumbly growl.

Nora the polar bear cub waits in quarantine soon after her arrival at the Oregon Zoo in September of 2016. *Photo by Shervin Hess for the Oregon Zoo.*

Nora walks around her enclosure at the Oregon Zoo in October of 2016. *Beth Nakamura © Oregonian Media Group.*

Nora eyes one of her favorite toys, a red ball, at the Oregon Zoo in March of 2017. *Stephanie Yao Long © Oregonian Media Group.*

Nora lunges after her red ball at the Oregon Zoo in March of 2017. *Stephanie Yao Long © Oregonian Media Group.*

Nora at the Oregon Zoo in August of 2017. *Dave Killen © Oregonian Media Group.*

Nora floats on a toy as she enjoys the pool during her last day on exhibit at the Oregon Zoo in September of 2017. *Dave Killen © Oregonian Media Group.*

An aerial view of Wales, Alaska, on April 2, 2017. Wales has a population of about 160 and is located at the westernmost point of the United States mainland, about sixty miles from Russia across the Bering Strait.
Dave Killen © Oregonian Media Group.

Gene Agnaboogok poses for a portrait outside his home in Wales in April of 2017.
Dave Killen © Oregonian Media Group.

A photo of his father that Gene Agnaboogok keeps at his house in Wales as seen in April of 2017. The photo shows Agnaboogok's father, Roland, atop a skin boat made from animal skins stretched over a wooden frame.
Dave Killen © Oregonian Media Group.

The remains of the last skin boat that was in use in Wales, Alaska, before they were completely replaced by aluminum-hulled vessels.
Dave Killen © Oregonian Media Group.

Gilbert Oxereok, a lifelong Wales resident and hunter, at the Wales multi-use center, in April of 2017.
Dave Killen © Oregonian Media Group.

Gene Agnaboogok lights a cigarette at his house in Wales, Alaska, on April 2, 2017. *Dave Killen © Oregonian Media Group.*

Whale bones protrude from the beach in Wales, Alaska, on March 31, 2017. *Dave Killen © Oregonian Media Group.*

A clip from the March 31, 1988, edition of *The Nome Nugget* shows the two polar bears rescued by Gene Agnaboogok near Wales, Alaska, one of whom would grow up to father Nora in 2015. *Photo courtesy of Diana Haecker and Sandra Madearis, for* The Nome Nugget.

Cindy Cupps,
Columbus Zoo and
Aquarium zookeeper.
Stephanie Yao Long ©
Oregonian Media Group.

Devon Sabo, Columbus Zoo
and Aquarium zookeeper.
Stephanie Yao Long ©
Oregonian Media Group.

Priya Bapodra, Columbus Zoo
and Aquarium veterinarian.
Stephanie Yao Long ©
Oregonian Media Group.

Shannon Morarity,
Columbus Zoo and
Aquarium assistant curator.
Stephanie Yao Long ©
Oregonian Media Group.

Karyn Rode measures a polar bear's vital signs during her field research on the Chukchi Sea off the northwest coast of Alaska in April of 2016.
Photo courtesy of Karyn Rode, U.S. Geological Survey.

Tasul, a female polar bear at the Oregon Zoo, sports a research collar that helped scientists study her behavior in July of 2013.
Benjamin Brink © Oregonian Media Group.

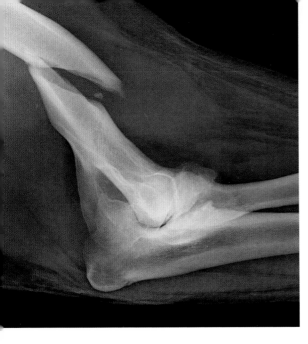

A radiograph shows a dramatic fracture Nora suffered in her humerus in early 2019.
Photo courtesy of Utah's Hogle Zoo.

Nora and Hope nuzzle with each other in the enclosure at Utah's Hogle Zoo, in Salt Lake City, in February of 2020.
Photo courtesy of Utah's Hogle Zoo.

Chapter 9

Tasul

At thirty-one, Tasul wasn't the oldest polar bear on record in 2016—the oldest had lived to forty-two—but she was one of the oldest still alive at the time. She had just lost her twin brother, Conrad. The two bears had been close. Tasul and Conrad sometimes slept curled up together, and when Tasul wanted to play, she bonked her brother with her head or batted a ball at him with her giant paw. Earlier in the year, the keepers had discovered that Conrad had an inoperable tumor; he had been euthanized, leaving Tasul as the sole polar bear at the Oregon Zoo.

Bears have long been noted for their ability to learn, with frequent starring roles in circuses and animal shows, where they've performed tasks ranging from roller-skating to balancing on balls to jumping rope. Those animals were often subjected to cruel training regimes and harsh punishment if they didn't perform. But wild bears, too, are capable of incredible feats of intelligence. Bears are thought to be among the most intelligent land mammals in North America and, in some cases, show cognition levels that rival those of primates. Grizzlies have incredible memories, often remembering the location of food sources for years or even decades. They can identify other individuals of their species and quickly place them in their complex social hierarchy, from up to two thousand feet away. Studies of sun bears, the smallest species, native to Southeast Asia, found that cubs are capable of copying the facial expressions of their contemporaries, a sign of social complexity and intelligence. Bears in national parks have such ingenuity that visitors constantly have to change up the type of "bear-proof" containers they use because the creatures learn new ways to break into them.

Among all eight species of bears, the polar bear is the only dedicated hunter. They are technically omnivorous, like their southern cousins—known to dine on berries, raid birds' nests, and rummage through trash dumps—but to satisfy their energy demands, polar bears need to consume an amount of fat obtainable only by stalking seals on open ice or plucking the unsuspecting pinnipeds from their breathing holes. That type of hunting takes patience, and polar bears will sometimes park themselves next to a seal's breathing hole for days, waiting for the creature to come up for air. It takes cunning, too. Rumors have circulated for years that, in an effort to conceal themselves while stalking prey, polar bears will use a paw to cover their nose, the only non-white part of their body, to blend in with the snow and ice. There's no evidence they actually do this—their sense of smell is their most powerful hunting tool, so it wouldn't make much sense to render it useless, not to mention the difficulty of hunting a seal on only three legs. Still, that the idea is even remotely believable is a testament to the intelligence of the polar bear.

Tasul was no different, holding in her brain thousands of years' worth of evolutionary adaptation. But her intelligence, combined with her size and power, meant she couldn't be tricked into anything. Portland zookeeper Nicole Nicassio-Hiskey knew a lot about animals before she met Tasul. She'd worked with Keiko, the orca made famous by the movie *Free Willy*, when she was a marine mammalogist at the Oregon Coast Aquarium, before heading north to tend to seals and sea lions at the Alaska SeaLife Center, on the Kenai Peninsula. She came to the Oregon Zoo in 2001 and, over the next nineteen years, worked with tigers, otters, and sun bears. She co-authored a book on animal behavior and how to curb destructive tendencies in pets by using enrichment to keep them from getting bored. She didn't pick favorites at the zoo, but her work with two of its most popular residents, polar bears Conrad and Tasul, was some of her most rewarding.

Establishing trust with Tasul had started simply. Nicassio-Hiskey fed the bear smelt and trout from her hands, talking to her softly from behind the bars. Tasul had learned to follow the keeper, who never forced the bear to take any action she didn't want to take; everything they did was on the bear's terms. Nicassio-Hiskey's biggest challenge was getting Tasul comfortable when keepers reached through the bars to touch her, which happened only as part of the bear's training regimen. It spooked Tasul at first—polar bears don't like to be touched. But Tasul learned to tolerate it through positive reinforcement from her keepers. Cooperation earned her papayas, bananas—peel and all—and squirts of fruit juice. She even got sherbet when her trainers were feeling generous. Over the years, Nicassio-Hiskey and Tasul built a rapport, the keeper learning to discern the animal's mood from subtle behaviors, the animal learning to tell what the human wanted from repetitive exercises in positive reinforcement.

The first time Nicassio-Hiskey held Tasul's paw, she was awestruck and humbled. It was the size of her head. She felt a deep sense of honor knowing that one of the world's fiercest predators would allow her so close. Nicassio-Hiskey remembers the date of their biggest breakthrough because it came on her birthday, December 4, 2011. It had been a rough stretch for the keeper.

She had just lost her father, and her emotions were raw, but her progress with Tasul had been encouraging.

A local company had custom-built a cage that attached to one of the dens, with a special cavity in the front big enough to accommodate Tasul's giant head. That morning, Nicassio-Hiskey stood in front of the cage, where she could hand-feed Tasul and monitor the bear's demeanor. Another keeper knelt in the back, where a door provided access to Tasul's hind paws.

As Nicassio-Hiskey occupied Tasul with carrots, the other keeper shaved the top of Tasul's paw and searched her skin for a vein. The fact that Tasul could be handled so easily was extraordinary.

And then the other keeper popped up.

"Oh my God," she said. "We got blood! We got blood!"

It marked the first time a polar bear had given blood without being tranquilized. Sedation is something keepers strive to avoid with zoo animals. Being at the business end of a tranquilizer dart is stressful for the animals, not to mention the keepers darting them, and anesthetization comes with its own medical risks. Tasul's willingness to give blood voluntarily meant her keepers could get samples from her whenever they wanted without putting her under. It opened up a whole range of research possibilities and turned Tasul into a bit of a celebrity in the small world of polar bear research.

Portland's zoo hadn't always been a bastion of research and conservation. Established in the late nineteenth century, it's the oldest zoo west of the Mississippi, and, like most zoos that have existed since before modern ideas of animal welfare came into fashion, its record is marked by tragedy and cruelty.

That history started with an English pharmacist named Richard Knight. Knight had been collecting animals from his seafaring friends and keeping them in the back of his downtown Portland drugstore, close to the docks on the nearby Willamette River. His collection, which included parakeets, monkeys, and two bears he kept in a vacant lot next to his shop, had begun to exceed the space available, so he began looking to offload some of his makeshift menagerie. He was hoping to sell the bears, "one young male brown, and a she-grizzly," he wrote to the mayor in June 1888. "They are gentle, easily cared for, and cost but a trifle to keep, and knowing they would prove a great source of attraction to the city park, would like an offer for them before sending elsewhere."

The city wasn't interested in paying for the bears and instead offered Knight two circus cages and permission to keep the animals in what would become Washington Park, tucked into the city's wooded West Hills. Care for the bears was still Knight's responsibility, at least for a few months, until he likely grew tired of having to feed them every day. He offered to donate the grizzly—it's unclear what became of the other bear—and the mayor accepted. In November, the first incarnation of what would become the Oregon Zoo was born.

By 1894, just six years later, the collection had grown to roughly three hundred animals. Just after the turn of the century, the zoo got its first polar bear. The animal, named Polar, was given to the zoo after a run in the Lewis and Clark Centennial Exposition. Polar was kept in a cage no larger than the average family room, constructed of cement blocks and iron bars, with a fence of two-by-fours and some chicken wire to keep spectators at a distance. Fed a gallon of milk and five pounds of cod liver oil per day, the bear didn't appear to be happy in his confined surroundings. "In his iron-barred cage, he stood peering into the open fields beyond," the local newspaper wrote at the time. "Swinging his head to and fro, to and fro unceasingly, it seemed he was longing for liberty."

The mayor, Harry Lane, visited the zoo and was disgusted. "I am of the opinion that there is nothing more cruel in the world than holding wild animals in captivity as is being done at the City Park," Lane told *The Oregonian*. "It makes me sick to think of it." A heated debate erupted among the city's residents, mostly focused on the well-being of Polar. The Oregon Humane Society said the bear should be put out of its misery, euthanized with chloroform to end its suffering. Letters to the editor poured into *The Oregonian*, pleading with city leaders to move the bear somewhere more hospitable. The mayor, who made no secret of his disdain for the zoo, vowed he would never fund the purchase of any animals from climates significantly different from that of Portland. Politicians come and go, however, and not everyone on the city council felt the same way Lane did. Lane would go on to become a U.S. senator, and Polar would live out his days in the concrete-and-iron cage in the park. After the bear died, in 1915, his pelt was mounted and displayed at City Hall. The polar bear enclosure was torn down, but you can still see remnants of it in Washington Park, the ruins of one wall next to a dirt walking path, just barely visible among a grove of rhododendrons and ferns.

It would be nearly thirty years before another polar bear came to the zoo. In 1943, a Soviet freighter docked in Portland, and on board was a fifty-pound male cub named Mishka, his fur matted and greasy from playing with the ship's gears and winches. The zoo's director convinced the captain to hand over the animal, and Mishka was placed in the zoo's main building, alongside snakes, cougars, monkeys, and at least one lion. By 1948 Mishka had grown to eight hundred pounds, and one evening, just before the zoo was set to close, he broke the padlock off his cage.

The first keeper to spot the bear was chased into a basement. Mishka then charged a car that was attempting to herd him back into the building. Pictures appeared in the local paper of the zoo director at the top of a ladder, hiding from the bear atop a twelve-foot cage. Firemen used hoses and chemical fire extinguishers in an unsuccessful attempt to force Mishka back toward his enclosure. Ultimately, it wasn't the efforts of keepers or firemen who got the massive bear back into his cage, but the temptation of food: He was lured back by a bit of horsemeat. Mishka lived at the zoo for almost another decade without incident before he was traded to the Dallas Zoo for a half dozen prairie dogs, a black leopard, and a ringtailed cat.

In 1959, the zoo moved to its current location, about a mile to the southwest of where Polar had lived out his days. The new facility was a vast improvement over its predecessor, for both visitors and the animals that lived there. The polar bear exhibit was surrounded by a deep moat and high walls from which zoogoers could look down on the animals. The enclosure featured a series of concrete platforms at differing heights, multiple pools, and dens where the bears could escape the public eye, should they desire. The first inhabitants of the new and improved habitat were Zero and Zerex, twin bears that had been born in the wild near Kotzebue, Alaska, about 175 miles northeast of Wales.

Within months, the five-hundred-pound bears had thrashed their new home, tearing four solid steel doors from their tracks. The nonfunctioning doors made it unsafe to clean the enclosure, so keepers had to sedate the bears with tranquilizers tucked into pieces of fish. A vet at the zoo said it took enough tranquilizers to knock out ten buffalo or give seventy-five people a good snooze. A zoo attendant stood by with a rifle while the repairs were made, just in case. Zero and Zerex would remain among the zoo's top attractions until they both died, in separate incidents, in 1966. First, Zero swallowed a rubber ball thrown into the pit by a zoo-goer. A necropsy revealed that the ball had blocked his intestines. Just a few months later, a gardener fell into the grotto while trimming some hedges. Zerex dragged the man around the enclosure for fifteen minutes, biting the employee's legs, hips, and arms and causing serious injuries. Keepers tried to ward the bear off with a fire hose, but when that didn't work, Zerex was fatally shot.

It didn't take long for the zoo to repopulate the polar bear grotto. The next year, two cubs from the Russian Arctic were purchased from a Dutch animal dealer for $2,200, about the price of a cheap car at the time. The bears were a gift from two local companies, ESCO Corp. and Hyster Co., which had naming rights for the animals. The male would be called Esco-Mo, the female Ice-Ter. The cubs were the same age, but they weren't related, and the zoo hoped to establish them as a breeding pair. In 1973, Ice-Ter gave birth to twins, but one was stillborn and the other contracted pneumonia and died less than a week later. She gave birth to another pair of cubs the next year, and all appeared well as she tended to them in a secluded maternity den out of view of the public. Keepers kept tabs on the cubs by listening through a speaker system, but after about a month, the den went quiet. Entering a week later, keepers found that Ice-Ter had eaten the cubs—gruesome by human standards, but not uncommon in the wild. Finally, in 1978, Ice-Ter successfully birthed and reared a single cub, a male named Cheechako, meaning "tenderfoot" in Chinookan, a family of languages spoken by the Native people of the Pacific Northwest. Cheechako moved to Utah's Hogle Zoo, in Salt Lake City, to make room for more cubs to be born. Ice-Ter would give birth to three more cubs over the following years, which found permanent homes in Sacramento and Seoul.

The zoo lost both adult bears in 1982, victims of a parasite in tainted salmon. It was sad—especially because Ice-Ter was found to be pregnant—but it also presented an opportunity. Times were changing, and the Oregon Zoo was intent on keeping up with the newest trends in animal care and management. The deaths of Esco-Mo and Ice-Ter made room for the zoo to begin construction on a new polar bear exhibit, which would look more like the bear's natural environment.

The new $2.6 million habitat was unveiled in 1986. Visitors entered through a long, cave-like tunnel where the screams of excited children echoed off the gray faux-rock walls. The first glimpse of the habitat came from floor-to-ceiling windows that gave zoo-goers an underwater view of the animals. There were no visible bars in the 0.4-acre enclosure, which was split into two public-facing yards connected by a door that keepers could close if they needed to. There was another yard, away from public view, that could be used if the bears needed privacy. The floors and walls were still concrete, but they were modeled to resemble nature—a far cry from the midcentury grotto, with its flat platforms and kidney-shaped pools. It wasn't the actual tundra, but there was grass, and big logs placed strategically for the bears to play on.

The zoo had come a long way from the concrete-and-iron cage that Mayor Harry Lane had railed against. When Tasul and Conrad arrived, in 1986, as two-year-old twins born in South Carolina, it felt like a turning point for the zoo, a new era in what had been a long and sometimes troubling history. Yugyan, a female who arrived at about the same time, lived at the zoo until 2008, when she died of kidney failure. Shivers, a male who came from the Columbus Zoo, sired a cub with Tasul, but it died just a few days later, and Shivers was transferred to a Mexican zoo in 1998. By the time Conrad died, he and Tasul had become mainstays at the zoo up on the hill, their longevity a testament to Portland's renewed commitment to animal welfare. The zoo's evolution mirrored that of similar facilities around the world, from little more than ill-kept menageries to institutions focused on education and conservation. The Oregon Zoo now hosts rescue and rehabilitation programs for a host of animals, from turtles to endangered butterflies to the California condor, which was barely pulled back from the brink of extinction.

It has also been on the leading edge of captive polar bear research.

Tasul's monumental blood draw in 2011 made headlines in Portland. When Karyn Rode, a wildlife biologist with the U.S. Geological Survey (USGS) living in the area, noticed the story in *The Oregonian,* she saw an opportunity to bolster her research. She'd worked with elephants and primates in the early years of her career, but most of her research had been on bears, specifically polar bears. She had spent years studying the animals on the ice off Alaska's west coast, where Nora's dad, Nanuq, had been orphaned.

She'd been using collars to track bears in the wild for years, but the devices, which used satellites to monitor the animals' location, only told Rode where the bears went. She wanted to know what they did when they got there. Tasul gave her that chance. Rode and another USGS polar bear researcher, Anthony Pagano, fitted Tasul with a collar-mounted sensor. It was something like an oversize Fitbit, with an activity monitor capable of measuring the bear's movements and determining if she was resting or active. The researchers took video of Tasul as she wore the collar, noting what she did and how she moved, matching their notes up to the data collected from the sensor. Now, when they attached the same sensors to wild bears, they would know how much time they spent resting and how much energy they spent moving around. Being able to catalog and analyze a wild bear's behavior would let researchers examine how that behavior changed as sea ice receded, got thinner, or vanished altogether.

The researchers also strapped a GoPro to Tasul's collar, which provided a bear's-eye view of what it was like to move around the enclosure. The footage was bumpy as Tasul loped about the yard, but there was novelty in seeing, from the bear's perspective, a toy fly from the hand of a keeper and Tasul lunging into the pool after it, thick white fur rippling in the water as bubbles of air streamed from her coat in slow motion.

The accelerometer and the camera would both prove their true worth on wild bears. Consumer GoPros couldn't stand up to an Arctic winter, Pagano learned the first year he tried, so he had special cameras designed to withstand temperatures of up to twenty degrees below zero. Over the course of three years, Pagano captured hours of footage from nine different bears. The cameras showed the bears swimming, resting, playing, and hunting. They told Pagano and other researchers exactly how frequently the bears ate; the best hunters snagged a seal every day or two. The cameras even caught parts of the mating process. Like Ian Stirling perched on the cliff at Devon Island, Pagano had found a way to see the elusive animals in their natural habitat, doing what bears do without human interference.

Tasul's ability to give blood regularly offered important insight into polar bears' diets, too. Working with Rode, Nicassio-Hiskey and the other keepers changed Tasul's diet, replicating what a bear would eat in the wild as closely as they could without access to actual seal meat.

Rode could then determine the degree to which chemical

components of Tasul's diet were reflected in her blood and hair. That allowed researchers to figure out what wild bears were eating by analyzing their blood, hair, and potential prey. Through months of training, Nicassio-Hiskey even got Tasul comfortable using a treadmill inside a metabolic chamber so Pagano could measure the energy demands of polar bear movement.

Tasul had been born in a zoo and had come to trust humans. She represented the best of what bears and people could accomplish when they worked together. For all the criticism levied at zoos—that the animals are unhappy or that the ends don't justify the means—the research Tasul participated in would have been impossible with wild bears. And the things Rode learned from Tasul allowed her to study those wild bears more effectively, indirectly benefiting all bears, both in zoos and out on the sea ice.

Tasul wasn't wild, and never would be, but she had spent her life with her brother, Conrad. When he died, she was alone for the first time. She was a bear who had been socialized to be around other bears and, though she was arthritic and a little slow, she was among the best-trained zoo bears on the planet, calm and wise. She was just the kind of animal who could serve as a mentor to a younger bear.

Chapter 10

Adaptation

Gilbert Oxereok knew he wanted to lead a hunting boat from the time he was young. He was only four or five the first time he went out on the Bering Strait, mostly because his parents didn't have a babysitter, but he remembers the older men trapping a beluga whale in a cove near Wales, killing the animal, and butchering it on a piece of nearby ice. As a kid, he spent his free time hunting birds with a slingshot one of his older brothers made for him. When he was twelve, he got a pump-action Winchester .22-caliber rifle in his Christmas stocking. From then on, if he wasn't in school, he was out on the hill behind Wales, hunting ptarmigan.

By his early teens, Oxereok had earned a shooting position on the boats that trawled the waters off the cape. The seating chart in a hunting boat ascends in rank from the stern to the bow, except for the captain, who usually runs the motor in the rear. In front of him sits the lowest rank, the coffee man. Ahead of the coffee man are hunters on either side of the boat, with the best shooters toward the bow. In the very front is the harpooner, charged with spearing any game killed by the other hunters so it won't sink or float away. Oxereok quickly came to know the ways of hunting from a boat—how to wait until he'd exhaled completely to pull the trigger, squeezing off his shot in that fraction of a second when the body is empty of air and completely still. He treats every shot like a math problem, timed to the undulating swell of the ocean: gauging the distance of the shot, the caliber of his rifle, the trajectory, and the wind, all in the brief window he has to size up his prey. He claims to be able to hit his mark from up to a mile away.

As a young man, Oxereok joined the military and spent a few years away from Wales, perfecting his marksmanship and learning to use a compass. But that was just to earn money to buy a boat. By the time he was in his early thirties, Oxereok was a captain.

Gene Agnaboogok got his first opportunity to join a hunt by chance, too. His dad was getting ready to join a crew that was going out onto the Bering Strait on a crisp and clear spring morning. There was an open seat on the boat, which one of Agnaboogok's older brothers would usually have taken, but none of them were home. Agnaboogok, then just six or seven years old, jumped at the chance to join the older men. The water was flat that day, and game—seals, walrus, and beluga whales—was abundant. The captain didn't have to go far from the shores of Wales for the five-man crew to encounter *ugruk*, bearded seals. Agnaboogok's memory of that day has faded over time, but he remembers what his role was. As the youngest, he was in charge of the coffee. Sitting near the rear of the boat, he made sure to keep the hunters' mugs full, all the time asking questions and trying to learn as much as he could through observation.

The ascent up the ranks of a hunting boat isn't swift. In the time that passed between his first day in the boat and the first shot he took at a seal, a few years later, Agnaboogok kept making coffee and asking questions. He learned that you never keep a weapon loaded while out on the water and that everything on the boat has its place. If you take something to use, you put it back where you got it. He learned how to haul *ugruk,* which can weigh up to seven hundred pounds, to a low, flat patch of ice and butcher it once he got there, separating the hide from the fat and the fat from the meat. Once the seal had been butchered, he learned how to keep the meat away from the boat's gas tanks so it didn't get contaminated.

He learned about the currents, which way the water was circulating as it pushed and pulled on the massive icebergs that flow between the Bering and Chukchi seas, and he learned to read the weather. His elders taught him to always keep one eye on the ocean and the other on one of several landmarks around Wales. Behind the village was Cape Mountain. Fairway Rock, a five-hundred-foot-tall islet that's home to migrating seabirds and not much else, sits about twenty miles off the coast. Northeast of that are the Diomedes, twin islands that straddle the International Date Line—the smaller one part of Alaska, the larger a Russian territory. Each of those landmarks, all visible from the waters around Wales, served as a weather beacon for hunters in the strait. When clouds formed above the landmasses—some in the village called them "wind caps"—that meant it was time to head home: Rough seas were coming.

Not everything the people of Wales hunt comes from the ocean. Moose and caribou and, in recent years, wolverines are known to walk the tundra around the cape. The polar bear is also prized, and a hunter's first kill is thought of as a rite of passage. Agnaboogok got his first bear when he was around twenty years old. It was just after Thanksgiving, and he had gone north from Wales on his snowmobile looking for game. Hunters rarely go out looking explicitly for polar bears; they are usually taken as prey of opportunity, tracked by their paw prints in the snow. Within a couple of miles Agnaboogok saw two bears, both more than ten feet in length, ambling on the sea ice. Polar bears are typically solitary animals, so it's likely that these two were a breeding pair, perhaps engaged in the long process of gaining trust prior to mating. The bears saw Agnaboogok just as he saw them and took off. He only had a moment to act and had fumbled his ammo. He ran back, gathered his shells, loaded his weapon, and aimed at one bear's hindquarters. Agnaboogok fired, and the bear crumpled. The other one continued to flee. Agnaboogok got off his snowmobile, approached the wounded animal, and killed it with a shot at the base of the neck.

Standing over the animal, Agnaboogok said a short prayer. He thanked the bear for giving its life and offered gratitude that the hunt had been a safe one. He butchered the animal out there on the ice, skinning the hide before cutting the meat into manageable pieces and loading them onto his sled. He motored back to the village and offered the meat to whoever wanted it. Agnaboogok knew the paws were coveted in Wales, so he gave two of them to elders back in the village. He saved the other two for his parents, who lived in the house up on the hill next to Oxereok.

Agnaboogok still lives in the house where he set out for his first seal-hunting expedition, where he delivered the paws of the first polar bear he killed, and where he brought the polar bear cubs after he rescued them from the collapsed den where he shot Nora's grandmother. It's a modular building crafted with efficient angles and a gabled roof that sits in a slight depression on the hillside, overlooking the Bering Strait. At one point the exterior walls may have been gray or green, but they've been beaten by the harsh winters into something in between, not quite one or the other. In 1988, when he discovered the cubs that would later be named Nanuq and Norton, it was a full house, Agnaboogok sharing the space—a living area with a kitchen in an alcove off to the side, two small bedrooms, and a bathroom—with his parents, a sister, her daughter, and another sibling's son.

Since then, his parents, Edna and Roland, have passed on and his siblings have moved out of the village. Agnaboogok now lives alone in the house. Inside, the faint smell of old fish and sea mammals mixes with the aroma of smoke from his woodstove. In the living room, most of the floor, chipped down to the plywood, is taken up by his mattress, piled with sheets and blankets. The house is warmed by an electric heater and the stove, which Agnaboogok feeds with driftwood when he has it or cardboard when he runs out. The straight black hair that's fallen around his shoulders since he was young is now streaked with gray. When he can afford cigarettes, smoke curls through the white goatee that circles his mouth and around his wire-rimmed glasses. The scar on his leg that the polar bear gave him has almost faded completely. He still hunts by snowmobile, but he's careful when climbing icebergs.

Wales doesn't have a sewer system, so most of the homes use a "honey pot"—a wooden box with a toilet seat that sits over a bucket—to dispose of their waste. When these get full, villagers put their contents in large black collection containers, and a few times a week, Agnaboogok hops on a four-wheeler, collects the contents of the black containers, and takes them to a small lagoon north of town for disposal. It's not glamorous work, but he doesn't mind it. When he's not working, he carves bones and ivory into intricate sculptures of birds and other animals, hoping to sell them to visitors passing through or in Nome, the biggest city in the region. When he's not hunting or working or carving, he hangs out with his cousin Josh Ongtowasruk, who plays guitar and sings songs in Inupiaq.

The landscape you can see from Agnaboogok's house doesn't look that different than it did when the missionaries arrived in the late 1800s or when the Spanish flu swept through in 1918. The slopes of Cape Mountain still drop down aggressively to the Bering Sea south of the village, the cliffs transitioning to beach as you move north toward the town. To the north, a big lagoon— five miles across in spots and seventeen miles long, with sandy inlets from the Bering Sea—still shimmers in the summer sun.

But the mission school is long gone, replaced by one of the newer buildings in town, which hosts classrooms, a large gymnasium, a library, a kitchen, and a bathroom with flush toilets and showers. The school serves not only the children of the village but those from surrounding communities as well. The rafters in the gym are decorated with banners from places like Teller, where the first reindeer camp was established in the missionary days; Shishmaref, the closest village to the north; and Little Diomede, the smaller (American) island in the middle of the strait. The gym sometimes stays open late for adult basketball games, and the building serves as a social gathering place almost as often as it does an educational facility.

Just north of the school is the Wales Native Store, one of two places in town to buy food, gas, ammunition, and other supplies. Almost everything on offer comes by plane, and the prices reflect the effort it takes to import goods to the tip of the North American continent. Next to the store, a row of houses stretches north along the village's main strip, Kingkinkgin Road. On the beach, a half dozen bowhead whale jawbones are planted in the earth, sun-bleached and pointing toward the sky, in front of a drying rack built of driftwood. The circular foundation of the old community center, a dome-shaped building erected in the 1980s, sits facing the sea. The dome once acted as a modern-day *qargi*, housing the tribal offices and hosting bingo and drumming nights. Kingkinkgin Road curves past the power plant, and next to that, under the shadow of a satellite tower, sits a depression in the earth, framed by two pieces of wood leaning against each other, with scattered whale bones arranged around the perimeter. These are the remnants of the sod house where Agnaboogok's mother was raised.

Past the power plant, a bridge carries the road over the creek as it washes into the Bering Strait. The road winds past the other store in the village, toward the post office, where mail and Amazon packages get delivered to P.O. boxes. There are no roads that lead to Wales, and it can be a difficult place to get to. It's gotten easier since the village got an airstrip, but the small planes that are supposed to land at the cape on a daily basis are often grounded by weather for days and sometimes weeks at a time.

Next to the post office is the multi-use center, referred to by everyone in town as the "Multi." When the dome was torn down, the Multi became the home of the tribal council offices, a couple of rooms for administrative workers and a communal space where official business is conducted and church services are held. Some nights, the halls echo with the muted rattle of bingo balls being tossed around as older residents mark up their cards. Other nights, the sounds of drumming and singing waft out of the Multi, pierced by the high-pitched giggles of toddlers who scamper up and down the hall. A bulletin board at the front is plastered with notices for jobs, classified ads for snowmobile parts, and posters warning of the dangers of alcohol and substance abuse, the lasting colonial legacy of the introduction of booze to the region by white men centuries ago. Like many Indigenous people, the Inupiat and other Alaskan Native groups have been subject to pernicious stereotypes about alcoholism. The village voted to ban the sale and importation of alcohol in 1981, and the law is printed out and posted on the wall of the Multi, just outside the tribal office. Oxereok works in the village as a lay pastor, and many of the problems people come to him with are related to depression and sadness, but rarely alcohol. Every once in a while, booze will make its way into the community, he says, either brought on a plane from Nome or schlepped on a snowmobile from a neighboring village, but drinking is relatively rare. "We're lucky we're not on the road system," Oxereok said.

North of town lie the airstrip and the cemetery. On the tallest dune stands the large cross erected after the Spanish flu swept through, its white paint chipped and fading. The population never recovered after the great sickness of 1918–19—though it had once boasted numbers of up to seven hundred people, it's hovered between 150 and 160 ever since.

It would be easy to look at Wales as a tragic place, marked by its remote location and forces beyond the control of the villagers who live there. Writers often describe the Arctic as desolate and barren, but this ignores the people who make it their home. There is a close sense of community in Wales, built on generations of self-reliance, and while the village is remote, it isn't isolated. Many homes have satellite TV and internet connections. When they don't, kids head to the school to log on to Facebook or Netflix, sitting on the gym bleachers as they wait for shows to download to their smartphones and tablets. On the court, teens sport Golden State Warriors gear as they work on their jump shots against a backdrop of rap music blaring from a Bluetooth speaker. White earbuds dangle from the ears of residents as they bump along the roads on four-wheelers.

The town can look very different depending on the season. In the summer, the hills behind the village and the plains to the north are green and vibrant, carpeted with tundra moss, Arctic grasses, willows, and berry bushes. Normal temperatures hover in the fifties, and in mid-June the sun shines close to twenty-four hours a day, with just an hour of twilight between 3 and 4 A.M.

As summer turns to fall, the grasses turn from green to red to brown and temperatures drop. By late October, the sun doesn't come up until late morning, and children with light-up sneakers blink along as they walk to school in darkness. The small lagoon behind the village is covered by a thin sheet of ice, and the tops of the mountains are capped with snow. Once winter fully takes hold, sunlight is hard to come by and snowdrifts pile up against the buildings, often up to their roofs. Historically, sea ice usually arrives by Thanksgiving, and at its peak it stretches far out into the Bering Strait. In the iciest years, locals say the frozen sea has stretched all the way to Little Diomede, twenty-five miles offshore. As spring breaks over the region, the days get longer, though the sea ice has historically clung to the shore until late May or early June.

Life in Wales has been defined by the seasons for as long as people have lived on the continent's most western cape. Spring is when hunters replenish their stores after the dark winter. Summer is for picking berries, catching fish, and hunting birds. Fall is for stocking up, knowing the long, cold months are just ahead.

But the historical seasons that the Native people of Wales have depended on since their ancestors settled there—the cycles of wind and temperature and snow and rain that dictate when it's time to gather plants and when it's safe to go hunting—are changing. And people like Gene Agnaboogok and Gilbert Oxereok are being forced to adapt.

Chapter 11

Arrival

Nora arrived in Portland in mid-September 2016. After a thirty-day medical quarantine, Nicole Nicassio-Hiskey and the other keepers orchestrated a series of introductions known in the zoo world as "howdies." First the zoo staff positioned the bears so they could see each other. Then Nora and Tasul were put in adjacent rooms, separated by only a metal screen. By October, Nora was nearly two hundred pounds and almost a year old. She didn't seem particularly interested in interacting with Tasul through the screen, but she didn't seem scared, either, and the keepers knew that socialization was key to her development.

A couple of weeks later, the keepers decided it was time to put them together.

Nicassio-Hiskey watched from the roof of the bear building, radio in hand. Catwalks rimmed the enclosures, giving her a bird's-eye view. She was charged with reporting, every twenty seconds or so, exactly what the bears were doing.

At least ten other keepers and veterinarians were stationed around the enclosure at every door, in case they needed to close one in a hurry. The exhibit was set up to limit dead ends, so Nora couldn't be backed into a corner. Keepers had frozen oranges and grapefruit they could toss to distract the bears. They had powerful hoses and fire extinguishers to separate the bears if they needed to. They cleared a channel so the constant traffic on the zoo's frequencies wouldn't prevent them from communicating.

They also had instructions: *Don't intervene unless you see blood.*

Polar bears usually growl or hunch their shoulders before their tempers flare, and Nicassio-Hiskey knew the kinds of things that might trigger a reaction in Tasul. The keepers gave Nora access to the main part of the exhibit, while Tasul was confined to the den.

After a few minutes, the door slid open and Tasul walked out. Nicassio-Hiskey's voice crackled over the radio.

As soon as the elder bear saw Nora, she broke into a sprint straight at the cub. Nora, confronted with an unfamiliar animal more than twice her size coming right at her, spooked.

She turned and ran.

Nicassio-Hiskey's radio fell silent. This wasn't going as any of them had hoped.

Nora darted over a log and through a tunnel that separated the two yards. Nicassio-Hiskey watched from the catwalk at the top of the fake-rock walls as the old bear, Tasul, lumbered behind.

Nora jumped into the pool, and the zookeepers held their breath. It was the only part of the enclosure where she could be cornered. For a moment, it looked as if Tasul might go in after her.

"T-bear!" the keepers called, fire hoses at the ready. The older bear backed off.

Animal introductions could be tricky, but Nicassio-Hiskey stayed calm. She had known Tasul for more than fifteen years, and she could tell when the bear was frightened, irritable, or aggressive. Tasul showed none of those signs as she followed Nora.

She was just curious. The older bear had gotten used to company. She'd been with her twin brother, Conrad, for years before he had died, two months earlier. Over the next few days, when the bears were together, Tasul tried to make herself approachable. She looked away when Nora got close. She lowered herself to the ground to appear smaller. She tried to entice Nora to play. But Nora wasn't interested.

The exhibit was closed to the public for the introductions, but the zoo released a video of Nora tentatively walking the yard, obviously wary of Tasul's presence. In the news release accompanying the video, the zoo described their first meeting as "extremely positive," but going in "slow motion." Behind the scenes, and unmentioned in the news release, Nora was starting to deteriorate emotionally.

Devon Sabo, the keeper who had watched Nora on the grainy red video when she was abandoned by her mom in Columbus, had accompanied the young bear on the trip from Ohio to Oregon. She spent nearly a week with the keepers in Portland, tutoring them on the finer points of Nora's personality and, more important, providing a familiar face to the bear as she got used to her new home.

Soon after Sabo left, though, Nora grew inconsolable. She barked like an angry seal, loud enough to be heard outside the building. Not even her favorite toys and treats could pierce the fog. She'd begun exhibiting stereotypic behaviors, actions repeated over and over with no particular purpose. She pawed at the concrete, digging imaginary holes. She paced in circles, bumping into toys but ignoring them. She fixated on her keepers, and any time they left the room, she threw a tantrum. She chewed on the bars of the crate she'd traveled in, making the same whirring noise she'd made when she nursed back in Ohio. Her symptoms got worse as the meetings with Tasul continued.

After one of her sessions with the older bear, Nora panicked and walked in circles for hours. Even when the bears were apart, her keepers sensed that Nora was apprehensive, as if she thought the older bear might be lurking.

And she wasn't getting better.

At the heart of Nora's story is a complicated question: Are zoos helping animals or hurting them?

On one end of the spectrum are animal rights absolutists, sometimes called animal liberationists. Some believe that humans and animals should be afforded the same rights, that personhood should go beyond humans to all manner of sentient beings, and that the treatment of animals as a natural resource is fundamentally flawed. This stance argues that utilitarian ethics—according to which the right thing is the thing that brings the greatest benefit to the largest group—demands that, because animals can suffer, their welfare must be taken into account. To exclude them from the calculus, they argue, is a form of discrimination called speciesism. Common among the groups at this end of the spectrum is a hard stance against animal captivity, especially captivity in the name of profit.

A 2003 study from the University of Oxford found that some animals—specifically lions, tigers, cheetahs, and polar bears— fared especially poorly in zoo settings. These large carnivores are often housed in enclosures that represent just a tiny fraction of their natural range, and confinement to such a small space (not the inability to hunt, as was previously thought) led to behavioral problems like pacing, as well as low infant survival rates. Polar bears had it as bad as any of the large carnivores, the study's authors found, with the average zoo habitat encompassing just a millionth of their range in the wild. The polar bears they examined for the study paced for roughly a quarter of their day and had infant mortality rates of 65 percent. (Michael Hutchins, then director of conservation science for the Association of Zoos and Aquariums, pushed back against the findings, telling *The New York Times* that the study included "broad generalizations" and "real weaknesses," adding, "I haven't seen any evidence of pacing" at the facilities included in the study.)

Even the animals that are content in their enclosures face a life as a spectacle. The zoos in Oregon and Columbus offered areas where polar bears could escape the public eye, but their habitats existed to showcase them. When they were in front of the public, their view was of an endlessly shuffling carousel of hairless creatures standing on two legs, all staring at them, some of them holding black boxes that flashed lights in their eyes.

The charismatic species—bears and lions and elephants among them—receive outsize attention, which critics say boils down to little more than a popularity contest. Those are the animals that get people through the gates, critics argue, and their conservation is given priority over that of other animals.

Zoos and their proponents trumpet the care the animals receive and point out that the wild, where zoo critics say captive animals should be allowed to remain, is not what the public imagines it to be. True wilderness, they argue, exists in an ever-shrinking supply, eaten away at the edges by human encroachment or wiped away in huge swaths in the name of development and capital. Few animals live as nature intended anymore. Not the rhinos, who are butchered for their horns, or the elephants, who are incessantly poached, or the orangutans, who have lost their homes to palm oil plantations. Not the manatees, the poison dart frogs, or even our own pets, which we domesticated into docility. Even the places we consider most wild, like the Arctic, have been home to humans since humans made homes. Every square inch of the planet is smudged with human fingerprints.

In the wild, whatever is left of it, polar bears roam hundreds of miles in search of food, but in zoos they don't have to. Zoo bears' diets are balanced and provided with a consistency they would never see in their natural environments. Whatever ailments they suffer, from toothaches to arthritis, are promptly treated by teams of professional veterinarians. When an affliction is beyond their realm of expertise, they call in specialists, often from across the country, to care for the animals. Regular meals and consistent medical care, defenders of zoos point out, translate to longer life spans for many captive animals than their wild counterparts enjoy.

But the benefits of zoos extend well beyond their walls. Zoos promote their conservation efforts through prominent displays around their facilities and on their websites. The Columbus Zoo partners with outside organizations to restore coral reefs in the Caribbean and funds vets to care for gorillas in Rwanda and the Democratic Republic of the Congo. About ninety miles southeast of Columbus, the zoo converted a former mining site into a nearly ten-thousand-acre open-range conservation park, where rhinos and cheetahs wander as they would in the wild. At the Oregon Zoo, in addition to the type of research Karyn Rode did with Tasul, there is a focus on preserving local species. Threatened silverspot butterflies are reared in a lab at the zoo before they are released on the Oregon coast in numbers ranging up to two thousand individuals. The zoo does similar work with western pond turtles, California condors, Oregon spotted frogs, and pygmy rabbits native to the nearby Columbia River basin. When critics accuse zoos of promoting only their charismatic animals—the bears and tigers and elephants that bring people through the gate—zoos say it's the draw of those high-profile species that funds the recovery efforts of butterflies and turtles and rabbits. A zoo full of frogs could never generate enough ticket sales to pay for the research into why frogs are disappearing in the wild. To save frogs, zoos need animals like Nora.

Even despite all their touted benefits, criticism from animal rights groups keeps zoos on the defensive. Keepers and PR staff often refer to "enclosures," "habitats," or "exhibits," but not "cages." Media access behind the scenes, when granted, usually comes with a caveat banning photography, to prevent the public from seeing bars. PR staff at zoos rarely release information when animals are faring poorly. Nora's physical ailments were not revealed to her adoring fans when they were discovered. When she was in the midst of treatment for her skeletal issues in Columbus, one of Nora's keepers was asked how the cub was doing and she responded, "Great." Metabolic bone disease never made it into press releases about her milestones or into videos of her cute antics. It's anyone's guess how many other zoo animals suffer through similar problems.

In general, zookeepers want what's best for their animals. Their jobs aren't high paying, the hours are long, and the conditions can be dangerous. They rarely have much say in the larger decisions that are made on behalf of the animals, such as which zoos they are sent to. But even with those challenges, keepers will tell you they form strong emotional bonds with the animals they care for. It's evident in the excitement with which Nicassio-Hiskey talks about Tasul's accomplishments and when Shannon Morarity's voice cracks as she describes caring for Nora as a week-old cub. It's evident in the sacrifices they make, like Priya Bapodra's canceled birthday plans or the Thanksgiving dinner she ate on the floor of the vet hospital as she helped care for an ailing young polar bear.

But the dedication and sacrifice of zookeepers can't change the fact that some creatures just don't do well outside of their natural environment. Animals like Nora present an impossible predicament. In the wild, she almost certainly would have died soon after her mom left her alone. Without human intervention back in Ohio, she likely wouldn't have survived the day, but that intervention had left scars on her psyche.

Born to a polar bear named Snowball in 1985 at the Toledo Zoo, in northern Ohio, Gus lived out his first few years in relative anonymity. That all changed when, about three years later, he moved to New York, in theory to breed, but in practice to become the face of the Central Park Zoo. Fresh off a $35 million renovation, the zoo boasted a snow monkey island where, in the winter, the resident Japanese macaques could leap from their icy surroundings into 104-degree hot tubs made of faux rock. It had a revamped tropical zone where strikingly green emerald tree boas slithered and black-and-white ruffed lemurs swung in front of walls hand-painted to look like the jungle. But Gus was the zoo's undisputed star. He was popular for all the reasons polar bears are popular everywhere. He was gregarious and charismatic, big and powerful, even if his pigeon-toed gait made him look a little dopey.

Gus's enclosure was typical of urban zoos. It wasn't huge, but it was painted to resemble the tundra, with areas of grass and gravel and a large, deep pool. Sometime in the early to mid-nineties, Gus started swimming laps. He would plant one of his hind paws on a fake rock, push off, then glide effortlessly across the pool before executing a diving flip turn at the window, where children gawked at the underwater view of the seven-hundred-pound bear. Then he'd swim back to the rock and repeat the whole process. Over and over, Gus splashed across the pool, sometimes for twelve hours a day.

His quirky behavior only added to his fame, but his keepers were worried. Something wasn't right with Gus.

Gus was exhibiting stereotypy, or stereotypic behavior, just like Nora would two decades later. It's not unique to animals—humans do it, too—but in any species it is at best a sign of stress. At worst, it can be an indicator of mental illness or a neurological disorder.

Philosophers and researchers have been arguing over animals' intellectual capacity for centuries. In the 1600s, René Descartes argued that animals were essentially automatons, programmed only to eat and reproduce, unable to reason or feel pain. About a hundred years later, David Hume proclaimed, "No truth appears to me more evident, than that beasts are endow'd with thought and reason as well as men." William Lauder Lindsay, a nineteenth-century physician and botanist from Scotland, thought that, in their essence, human and animal brains were the same. Lindsay's ideas weren't as progressive as they sound—he also believed that the mentally ill, criminals, and non-European people could be classified in the same category as nonhuman animals—but he's remembered for another contribution. He believed other species could suffer from a number of the same mental ailments as people, writing that many animals had the capacity to be afflicted by illnesses ranging from nymphomania to delusions to dementia to melancholia.

Charles Darwin shared similar views about the intelligence of animals, and his work laid the path for a new field of study that would come to be known as ethology, the examination of animal behavior. Working in animals' natural habitats, ethologists have found complex social communication in whale songs, altruism in bonobos, and evidence of grief in baboons. The hallmark of higher intelligence, though, remained self-awareness. In 1970, Gordon Gallup Jr., a psychologist at Tulane University, decided to find out if animals could recognize themselves. Gallup introduced a group of four chimps—two male and two female—to a mirror. At first the chimps responded with threatening behavior, presumably because they saw their own image as a rival. As they got closer to their reflections, however, the chimps began grooming parts of their bodies previously out of their field of vision. They made faces and picked their noses. It seemed as if the chimps knew they were looking at themselves, but Gallup wanted further proof. He anesthetized the animals and made markings on their brows and ears with an odorless dye. When they woke up, they were given access to the mirror, and the amount of time they spent looking at themselves increased, as did the number of times they touched the areas with the dye. They also turned their bodies to get a better view. Gallup took this all to mean that the chimps recognized that the images in the mirror were of themselves. The mirror test, as it would come to be known, would be repeated often, across dozens of species. Elephants, orcas, bottlenose dolphins, and a number of the great apes—bonobos, orangutans, and gorillas—have shown at least some evidence of self-recognition. Even some ants, when marked with dye, showed behavioral differences when put in front of their own reflection. Though the mirror test has been criticized as unreliable and prone to false positives, it is still widely thought of as the best indicator of self-awareness in animals.

As the environmental movement gained momentum in the seventies and eighties, and the animal rights movement with it, Descartes's opinion that animals are little more than organic robots was widely viewed as discredited. In 2012, a group of neuroscientists gathered to make it official. The Cambridge Declaration on Consciousness stated directly that "the weight of evidence indicates that humans are not unique in possessing the neurological substrates that generate consciousness. Non-human animals, including all mammals and birds, and many other creatures, including octopuses, also possess these neurological substrates." The question was no longer *Do animals have emotions?* but rather *What kind of emotions do they have?*

In humans, an elevated heart rate and dilated pupils might be a sign of anxiety, but there's no way to know if animals experience emotions like anxiety the same way we do. As scientists sought to understand the inner workings of the animal mind, they began looking for new ways to talk about the psychological well-being of animals, some using the term "affective states" to describe a combination of their physiological, biological, and behavioral reactions to their environment. Researchers looked at heart rates and the levels of certain stress hormones like glucocorticoid. They checked for lameness, for weight loss, and for overall health. Finally, they looked at the animal's behavior—whether the creature was acting within the norms for that species. They sought to differentiate between momentary stress, like being chased by a predator, and an animal's mood over longer periods of time. Taken together, these observations could give caregivers—whether they were zookeepers, beef ranchers, lab technicians, or pet owners—a better idea of the mental welfare of an individual creature.

But sometimes behavior that looks like it's caused by mental illness can be the result of something else entirely. In her book *Animal Madness,* historian of science Laurel Braitman relates the tale of a troubled tree kangaroo at the San Antonio Zoo who had been repeatedly attacking her newborn joeys. Every time a keeper entered the enclosure, the mother batted the tiny creatures around with her paws, clawing at the vulnerable infants. The zoo feared the mother was suffering from a mental break and called in veterinary specialist Mel Richardson. When he approached the kangaroo, she ran to her young and started pummeling them. When he stepped back, though, she stopped. He repeated the process and, sure enough, she again started hitting the youngsters when he got close but stopped when he retreated. "I realized that she wasn't viciously attacking her babies at all," Richardson told Braitman later. "She was trying to pick them up off the floor, but her little paws weren't meant for that. In her native Australia and Papua New Guinea her babies never would have been on the ground. Her whole family would have been up in the trees." The mother wasn't attacking her joeys; she was attempting to protect them from the people intruding on her space. The zoo made changes to the kangaroos' habitat, moving the nesting area up and farther from the door, and the troubling behavior stopped. Abnormal behavior can't always be curbed with changes to an animal's environment, though, and zookeepers often turn to pharmaceuticals when their animals start acting abnormally.

Most mood-altering drugs were tested on animals before they got anywhere close to humans. Chlorpromazine, later sold as Thorazine, was first developed as an antihistamine in the early 1950s, but researchers found it had a depressant effect on rats treated with the drug. Marketed as the world's first antipsychotic, chlorpromazine was widely used on humans within a few years of those early tests on rats. Meprobamate, marketed as Miltown, was a powerful drug that first showed promise in tests on rhesus monkeys around the same time. The development of meprobamate and chlorpromazine paved the way for an influx of antidepressants like Zoloft and Prozac, as well as today's common tranquilizers, benzodiazepines like Valium and Xanax. Eventually the drugs would come full circle: Animals, once used only as test subjects, would become patients. One of the first was a gorilla named Willie B.

Born wild in Africa, Willie B. had been captured in the early 1960s and sent to Zoo Atlanta. Over the winter of 1970–71, Willie broke a window in his enclosure and had to be moved to a much smaller cage for six months while the glass was replaced with heavy metal bars. Looking to make his stay in the tiny enclosure more tolerable, Willie's keepers mixed a dose of Thorazine in with the soda he drank every morning. Mel Richardson, the same vet who helped diagnose the tree kangaroo in San Antonio, was working at the zoo at the time. He said Willie reacted to the antipsychotic in a similar way to how a human patient would, shuffling about with glazed-over eyes. "It was a little like watching the men in *One Flew Over the Cuckoo's Nest*," Richardson told Braitman, "except Willie was a gorilla."

Today, mood-altering drugs are used in zoos across the country and around the globe. SeaWorld dosed its California sea lions with Haldol, an antipsychotic. A young female walrus at Six Flags Marine World, in Northern California, was given antipsychotics after she was observed regurgitating her food repeatedly. At the zoo in Toledo, Ohio, keepers have used antidepressants, tranquilizers, and antipsychotics on zebras, wildebeests, ostriches, and a swamp monkey. In one case, a female gorilla named Johari was given Prozac to ease her aggression, which grew worse just before she had her period. "[Psychotropic drugs are] definitely a wonderful management tool, and that's how we look at them," the zoo's mammal curator, Randi Meyerson, told the local paper. "To be able to just take the edge off puts us a little more at ease." The market for psychotropic drugs in the animal kingdom has grown to such an extent that Prozac is now available in a number of animal-friendly flavors like anchovy, beef, liver, fish, double fish, and triple fish.

But pharmaceutical remedies only provide a Band-Aid for the underlying causes of captive animal distress, one of which is the tedium of spending long periods of time confined to a small space. The challenge for keepers of smart animals—like gorillas, elephants, and polar bears—is keeping their minds engaged.

Gus's keepers were never really sure what triggered his stereotypic lap-swimming behavior, but there are some clues. In his early days at the zoo, one of Gus's favorite activities was stalking children next to the window that looked into his pool. "He liked to see them scream and run in terror—it was a game," the zoo's animal supervisor told *New York* magazine in 1995. "But we didn't want heart attacks, so we put up barriers to keep people farther from the glass." Soon after the barriers went up, Gus began swimming the laps that would make him famous.

The first stories about Gus's stereotypic behavior came from *Newsday*, but within days the bear had garnered headlines in *The New York Times* and David Letterman was cracking jokes about him on late-night TV. The Canadian band the Tragically Hip wrote a song called "What's Troubling Gus?" His behavior inspired children's books and a short play. New Yorkers embraced the bear as a symbol of their own collective anxiety. A zoo spokeswoman said Gus was a healthy bear, just one with a case of "mild neurosis." A *New York Times* columnist responded, "JUST A MILD NEUROSIS? CALL YOURSELF A NEW YORKER AND ALL YOU GOT IS MILD NEUROSIS?" Gus became known as the "bipolar bear," and calls flooded into the zoo inquiring about his condition. The zoo paid $25,000 to hire a behavior specialist named Tim Desmond, who had previously worked with the orca from *Free Willy*, and New Yorkers were further delighted that, like them, Gus also had an expensive therapist. His fame translated to a spike in ticket sales at the zoo, but his keepers were worried. "He's not meeting our criteria for quality of life," Desmond said. "We're trying to perfect his life style."

The diagnosis? Gus was bored. The prescription: a dose of

Prozac, which marked what is believed to be the first time a zoo animal was treated with the wildly popular antidepressant. Aside from the drugs, Gus just needed more stuff to do. To occupy his time, Desmond helped the zoo concoct a "play area" for the bear. He was given rubber trash cans with food inside to maul. His fish snacks were frozen into blocks of ice. His chicken was wrapped in rawhide. The obsessive swimming lessened, but it never completely went away.

In zoos, polar bears are among the animals furthest removed from their natural environment. Gus's enclosure, which he shared with two female bears, was state-of-the-art when he moved into it, but it was still only five thousand square feet, approximately 0.00009 percent of the area he would roam in the wild. Even after the expensive treatment, the games and puzzles, and the attention of millions, whether Gus was content or not remained up for debate.

In Portland, Nora's keepers would face a similar dilemma.

When her pacing and tantrums continued, Mitch Finnegan, an Oregon Zoo veterinarian, prescribed Nora alprazolam, better known as Xanax, to calm her. She took a pill once in the morning and another in the evening, hidden in ground horsemeat, for a total of four milligrams a day. Her mood improved as the medication took effect, but she remained anxious. Her dosage was upped to six milligrams per day.

Two weeks later, Nora still paced. The zoo called in an animal behavior specialist who recommended a different approach. In addition to the Xanax, Nora was put on fluoxetine, a generic version of the antidepressant Prozac. It was a sign that her mental recovery would take longer than originally expected.

Chapter 12

Sinking into the Sea

Off the coast of Alaska, the shadow of a helicopter swept across the snow. A hundred feet up, Karyn Rode scanned the craggy landscape for tracks.

Starting in the spring of 2008, Rode traveled to her remote research base on the Chukchi Sea, north of the Arctic Circle. A charter flight dropped her off in late March at a remote airstrip next to one of the world's largest zinc mines. In the summer, massive excavators pulled minerals from the earth, but in the winter the company that runs the mine, Red Dog, let researchers use its port, a couple of hours' drive west of the mine and about fifteen miles south of the Native village of Kivalina, as a jumping-off point for flights over the sea ice. The frozen sea looked like a whitewashed version of the moon. It always amazed Rode that anything could live out there.

The open terrain was Nora's ancestral homeland. Her father, Nanuq, was orphaned in 1988 when Gene Agnaboogok fell into the den about two hundred miles from here. Polar bears can roam farther than that in a week.

"Is that a track?" Rode's voice crackled over the headset as she guided the pilot from the passenger seat. "I think that's a track."

The paw prints faded out in blown snow.

The helicopter bobbed and lurched as they followed different tracks on the frozen sea, most of them leading nowhere. Eventually Rode spotted a bear lumbering across flat, open ice.

"You see it right in front of us there?" the pilot asked over the headset.

"I do."

Rode knew you couldn't just dart every bear you saw. If it was a female with a cub, there was a chance the youngster could run off and get lost, which for a young cub would likely mean death. She also wouldn't dart a bear that was near broken ice or open ocean. If it ran into the water or broke through the ice before it was fully sedated, it could drown. She looked for healthy bears alone on large expanses of open, flat ice. The animal in front of her checked all the boxes. The helicopter swooped low just as the bear broke into a lope heading north. Another biologist stood in her seat, aimed a tranquilizer gun at the bear's shoulder, and fired a dart. Within minutes, the pilot gently landed the skids near the sleeping mountain of fur. Rode went to work.

The United States is home to two of the nineteen populations of polar bears: the Chukchi Sea bears, Nora's kin, who range up and down Alaska's west coast and across to Siberia, and the Southern Beaufort Sea bears, along Alaska's northern coast. In 2005, the Center for Biological Diversity, an environmental advocacy group, petitioned the government to grant them protection under the Endangered Species Act. At the time of the petition, little was known about the size of the Chukchi population. A crude estimate of population density had yielded a guess of 1,200 to 3,200 bears, but that data was nearly twenty years old and considered unreliable. There were more recent, and more reliable, numbers for the Southern Beaufort bears—and those numbers showed the population increasing.

The Center usually petitioned on behalf of animals like the spotted owl, which had seen its numbers decline as a result of human incursion into its habitat. That made the polar bear petition unusual. But that American polar bears weren't facing an immediate threat was not a fact the advocates tried to hide. "While most populations are currently reasonably healthy and the global population is not presently endangered," the petition read, "the species as a whole faces the likelihood of severe endangerment and possible extinction by the end of the century." Other environmental groups signed on to the petition, and the government eventually agreed to nominate the bears for protected status. Over the ensuing year, the government called for studies and analysis from scientists including Rode and Ian Stirling, which said that polar bears were likely to lose as much as 42 percent of their optimal habitat by 2050. The public was invited to weigh in, too, and did so with roughly 670,000 responses. Hearings were held in Washington, D.C., Anchorage, and Utqiagvik (formerly known as Barrow), on Alaska's north coast. Polar bear advocates would have to go back to court several times, but in May 2008, three years after the original petition had been filed, the government reached its decision: The polar bear was a threatened species. Though the federal government wouldn't admit it, the threat was us.

Under most circumstances when an animal is listed under the Endangered Species Act, experts come together, assess the risk, and develop a recovery plan. For spotted owls, the federal government designated nearly ten million acres as critical habitat for the birds, protecting the old-growth forests they relied on from logging. With polar bears, though, their entire habitat was at risk, and the threat wasn't any one thing. It was the economic system that relied on the burning of fossil fuels on a massive scale. It was the blanket of carbon that was keeping heat from escaping the atmosphere, melting the Arctic ice, shrinking the frozen hunting platforms the bears needed to survive. When the polar bear was granted federal protection, it was the first time an animal had been given that special status not owing to an imminent threat of extinction but for a threat that computer models predicted would spell their future demise.

Political reactions fell along largely predictable partisan lines, conservatives blasting the ruling as an overreach based on flawed science that sought to protect a species that appeared to be thriving, and progressives lamenting that the regulations didn't go further to protect the bears from a situation the science said would deteriorate rapidly.

On its face, the decision seemed like a victory for the environmental groups, but Interior Secretary Dirk Kempthorne made it clear that the listing of a species would not dictate American policy on carbon emissions. The Endangered Species Act "is not the right tool to set U.S. climate change policy," he said at a press conference. "This has been a difficult decision. But in light of the scientific record and the restraints of the inflexible law that guides me, I believe it was the only decision I could make."

Polar bears were already protected against commercial hunting by the Marine Mammal Protection Act and the Agreement on the Conservation of Polar Bears, both enacted in the early seventies, as well as by another treaty, signed by Russian and U.S. officials in 2006, that strengthened protections for the Chukchi bears. But listing the bears as threatened required the government to look for critical habitat to preserve, which it did two years later, and to draft a recovery plan. It also meant that, under the Endangered Species Act, federal agencies would have to "ensure that any actions they authorize, fund, or carry out are not likely to jeopardize the continued existence" of the bears "or result in the destruction or adverse modification of designated critical habitat."

The George W. Bush administration had been reluctant to act on climate change, even as the scientific consensus grew, and the listing of a bear wasn't going to change that. Along with the threatened status, the government also issued a special rule under section 4(d) of the Endangered Species Act. The "4(d) rule" allowed for flexibility under the act, giving the U.S. Fish and Wildlife Service discretion to permit gas and oil drilling in the bear's habitat because the drilling itself posed no direct threat to the animals. Not much changed after Barack Obama took office. While more progressive on environmental issues overall, the Obama administration couldn't "connect the dots," as the Endangered Species Act requires, between any one source of greenhouse gases and the threat to polar bears, so the rule stayed in place and fossil fuel extraction in the Arctic continued.

Rode's boots crunched on the ice as she hopped from the helicopter. She was about eighty miles off the west coast of Alaska, on the ice of the Chukchi Sea. First she took samples: blood, hair, stool, and a biopsy of fat. The bear groaned and twitched. She ran a tape measure along the bear's spine and around its midsection. She and another biologist set up a heavy-duty tripod with a net and a chain hoist. They rolled the bear onto the net and lifted it to measure its weight—542 pounds—then tattooed its inner lip so it could be tracked if it was ever caught again. This one was Bear 21736. It was a female, so Rode fitted it with a radio collar. (Male necks are too big for them.)

"Good looking bear," she wrote in her log.

She stuffed the test tubes into a box lined with hand warmers so they wouldn't freeze. Later, Rode would analyze the samples, based, at least in part, on what she had learned from Tasul at the

Oregon Zoo. The results would give her clues as to what the bear had eaten.

About an hour after Bear 21736 was sedated, she began to rouse, lifting her head in time to see the helicopter take off.

Rode began her yearly trips to the Chukchi the same year the polar bear was listed under the Endangered Species Act. Over the following decade, she gathered enough anecdotal evidence to know that the animals in that population appeared to be doing well. In 2019, Eric Regehr, who studied the demographics of the Chukchi population for the Fish and Wildlife Service, found the same thing. His was the first reliable population estimate for the Chukchi bears, a group of roughly three thousand robust and hearty animals.

Summer sea ice in the Chukchi had declined precipitously—the thirteen lowest measurements of summer sea ice in the satellite era had come in the last thirteen years, between 2007 and 2019—but the bears seemed to be doing fine. It didn't exactly square with the image put forth in the push to get the animals protected status.

A lot of that could be chalked up to the geography of the seafloor. Much of the Chukchi Sea is shallow, sitting over a continental shelf that extends out from the land on either side of the Bering Strait. These shallow waters support a robust and productive ecosystem teeming with life, from microscopic plankton to whales, seals, and polar bears. The bounty sitting just offshore is one of the reasons people living in Wales have been such successful hunters throughout history. It's also why the Chukchi bears might have an easier time finding food than some other populations even as ice diminishes. On the Chukchi, bears can follow the ice north and remain over shallow waters that support their main prey, the ringed seal. The same doesn't hold true for America's other population of polar bears, the Southern

Beaufort Sea group, who live around the corner from the Chuk-chi Sea on Alaska's north coast. The Beaufort Sea, on the south-ern edge of the Arctic Ocean, is cold and deep. When the sea ice begins to melt in the spring, the bears there must make a choice: Go with the ice over deep water, where seals are scarce, or stay on land and fast for the summer. Between 2000 and 2010, research-ers saw an alarming 40 percent drop in the Beaufort population. The bears came out of free fall and stabilized toward the end of the ten-year study, and researchers never pinned down the exact reason for the dip. But it was clear that the Beaufort bears oper-ated on a thinner margin than their cousins in the Chukchi.

And therein lay the problem. Each of the nineteen popula-tions of polar bears relies on its own local ecosystem and faces its own challenges. Some of those populations were likely declin-ing, but at least one was thought to be growing, and two others were considered stable. Twelve were too remote to even hazard a guess.

For those inclined to dismiss the science on climate change, or to question the polar bear's inclusion on the threatened spe-cies list, the areas where data is lacking make for easy targets. And the unknown areas may be growing as the ice shrinks.

The wide, flat areas Rode needed to sedate the bears were harder to find in 2017 than ever before. Many parts of the Chuk-chi were slush or open water. In a good year, Rode tagged more than thirty bears. That year, she caught three.

Ten days before her trip was to end, the ice ran out. For the first time, Rode and her team had to pack up early and leave. The next year, 2018, the entire research trip was canceled because of a lack of workable ice. The ice was worse in 2019. Bear 21736 may very well have been one of the last Rode would ever catch on the Chukchi.

Chapter 13

Alone Again

Tasul had been slowing down for months. In July of 2016, her keepers noticed she was favoring her rear right leg, swinging her other hind leg quickly to avoid ever putting her full weight on the sensitive limb. The limp was worse after her sessions on the treadmill. They knew she was arthritic and had degenerative joint disease in her right hip, the cartilage worn away from the joint after three decades of supporting five hundred pounds of fat and fur and muscle. It had been an issue for a while, but it appeared to be getting worse, so they cut back on her treadmill time. On a few occasions, they noticed dried blood near her tail and noted in her chart that she had recurring vaginitis. At one point keepers wrote that she had maggots on her genitals. They prescribed salt baths and an antiparasitic called ivermectin. In early October, keepers noticed green mucus collecting around Tasul's nostrils. She was acting strange, too. "It is usual for her to be slightly lethargic this time of year but she seems a little different to the keepers now," Mitch Finnegan, the zoo vet, wrote in her chart.

Finnegan put the snot sample under a microscope and saw that it was made up almost entirely of white blood cells. He ordered a full panel of bloodwork. Using the training she'd developed with Nicole Nicassio-Hiskey, Tasul voluntarily gave a sample a few days later. The results came back high in a number of concerning categories: Her hemoglobin concentration was above recommended levels, as were her triglycerides, cholesterol, and creatinine. She also showed elevated levels of blood urea nitrogen, a chemical waste product. The old bear's kidneys, the organs charged with cleaning and filtering her blood, weren't functioning properly. She also had what was likely a dental abscess, a respiratory tract infection, and a growth in her nose. They started flushing her system, encouraging her to drink as much water as possible, while replenishing her nutrients with an electrolyte mixture, essentially the polar bear version of Pedialyte.

She seemed to bounce back, for a few weeks at least. She was still showing signs of old age, to be sure, but with a restored level of energy. Then, about a month later, keepers noticed blood in Tasul's mucus. They ordered a thorough examination and quarantined the bears from each other. Just a few days short of Nora's birthday, in early November, keepers sedated Tasul and gave her an ultrasound. The imaging showed abnormal growths in her abdomen. Her bloodwork was slightly improved, but a biopsy confirmed early-stage kidney failure. They couldn't be sure, but the veterinarians also suspected the old bear had an ovarian tumor that could, if left untreated, spread cancer to other vital organs.

The introductions to Nora, which had failed to produce many encouraging results, were called off. Tasul needed surgery.

Early on November 17, Finnegan poked a dart filled with sedatives into Tasul's thick hide, and she went down in minutes. Together, the keepers rolled her onto a cargo net and lifted her into a zoo van for the short ride from the polar bear enclosure to the medical center. In the operating room, she was positioned on the operating table and draped with sterile cloths. Finnegan stood to the side as the surgeon made a nearly ten-inch incision lengthwise along her belly.

It was worse than they'd feared. Cancer lined the inner wall of her body cavity, from her pelvis to her kidneys. Her lymph nodes were inflamed.

She had likely been in a great deal of pain. Even if they could remove the diseased parts of her body, the recovery from an extensive surgery like that would be long and uncomfortable, and, at her age, the risk outweighed the benefit. Finnegan believed she couldn't be helped. Less than an hour into the procedure, he picked up the phone. He called the deputy director of the zoo and the curator. He called the keepers, including Nicassio-Hiskey, who was home with a bad case of the flu. After a few brief discussions, the path forward was clear. Finnegan helped administer a cocktail of pentobarbital to further sedate her, and potassium chloride to stop her heart. Tasul died within minutes, just a couple of weeks shy of her thirty-second birthday.

Afterward, Nicassio-Hiskey went to the zoo. Tasul lay on the operating table while her keepers cried and traded stories during a makeshift wake. Nicassio-Hiskey had been awestruck the first time she held Tasul's paw, more than a decade before, and she'd been humbled every time the bear allowed their interspecies contact. They could touch her as much as they wanted now.

The decision was heartbreaking but clear, and not entirely unexpected. Tasul had lived well beyond the median life span for a member of her species, wild or captive, and contributed more to polar bear research than arguably any other creature. The drugs that ended her life also alleviated whatever discomfort she'd been silently suffering through as tumors grew in her body and arthritis crippled her movement. The pain coming to an end was a blessing, Nicassio-Hiskey knew, but Tasul's death left a hole in the keeper's heart that she would not quickly fill.

It also left Nora alone again.

When Donald Trump won the 2016 election, nine days before Tasul's death, he became the first president-elect to openly reject even the idea of climate change, the logical result of a successful, decades-long misinformation campaign. At various times leading up to his candidacy, he called it a "hoax," "bullshit," and an elaborate ploy "created by and for the Chinese in order to make U.S. manufacturing non-competitive." Usually, his evidence for this far-reaching conspiracy was that there was cold weather somewhere. How could the planet be heating up, he wondered, if it was snowing in Texas? In January of 2016, the federal agencies he was campaigning to lead announced that the previous year had been the hottest ever recorded, but according to candidate Trump, there were differing opinions among scientists on both sides of the climate debate. And this was true, if you take a 97 percent consensus among experts to mean that the science was unsettled. He pushed for the rollback of environmental regulations across the board, the Environmental Protection Agency being his primary target. "Who's going to protect the environment?" an interviewer from Fox News asked him. "We'll be fine with the environment. We can leave a little bit [of the EPA], but you can't destroy businesses," Trump responded. He championed oil and fracked natural gas. He wanted to open up the Arctic National Wildlife Refuge to drilling. Coal, he told his supporters, would see a resurgence under a Trump administration.

Trump's style may have been unique, but the substance of his remarks only represented a continuation of the policies set by the Republican presidents who preceded him. Soon after his election, Ronald Reagan had appointed an energy secretary who didn't believe in climate change and tried to cut funding for Charles Keeling's carbon-dioxide-monitoring efforts at the Mauna Loa Observatory. George H. W. Bush showed some promise as a candidate, paying lip service to environmental issues on the campaign trail, but failed to establish any meaningful benchmarks on carbon dioxide emissions. George W. Bush declined to implement the Kyoto Protocol, the treaty signed by his predecessor that was meant to curb greenhouse gas emissions, because he said it would hurt the U.S. economy. He made no mention of the economic damage that would come to pass under even the most conservative models of climate change. Environmental advocates didn't spare Democratic presidents from criticism, either. Even flanked by Vice President Al Gore, one of the most outspoken mainstream politicians on environmental issues, Bill Clinton was blasted by critics for not going far enough on issues of endangered animals, public lands, and fuel efficiency. Barack Obama signed on to the Paris Climate Agreement of 2015, which was seen as a monumental achievement by some, but to others it was more of the middle-ground approach that had perched the world on the precipice. Confronted with evidence that we were approaching a point of no return, even progressive presidents often chose a path that would fail to avoid the worst consequences of global warming.

Trump's stance on the environment and climate change wasn't novel. He was echoing decades of talking points first promulgated by conservative think tanks and industry groups like the Global Climate Coalition. Where he differed from past Republican politicians—not just on environmental issues but on almost everything—was that he said the quiet parts out loud. Where previous presidents had at least said they cared about climate change, before taking little or no action to confront it, Trump plainly said he thought it was phony.

Winter brought snow to Oregon, and Nora was introduced to the people of Portland. Visitors saw Nora in an environment that bore a fleeting resemblance to her natural habitat. She romped through the yard as fat flakes fell around her. She pressed close to the glass as zoo staff engaged her in a game of peekaboo. In one of the coldest and snowiest winters the Pacific Northwest had ever seen, Nora's warm breath steamed the glass. After the crowds left, she interacted with the security guards, following them along the exhibit's periphery and splashing in the pool to get their attention. She put her nose to the glass, trying to smell their coffee. When custodians washed the windows with giant, furry cleaning mitts, Nora's paws mirrored them from the other side. Like Knut in Berlin and Gus in New York, Nora was a polar bear that fed off attention from, and proximity to, humans. Without an audience, she grew ornery. Before the crowds arrived, she stomped around the yard and chuffed in frustration to anyone who would listen. To Nicassio-Hiskey, it seemed as if Nora was complaining that no one had come to see her.

As soon as the first people showed up, the growls stopped and Nora belly-flopped into the pool, straight toward the windows. She was back to being the rambunctious cub that had given the residents of Ohio so much to swoon over, and Oregonians took to her in the same fashion. Hundreds crowded her exhibit on weekends, streams of toddlers shuffling up to the glass, their shrieks echoing off the faded rock walls. In Columbus, her antics had endeared her to millions on the internet, and the same held true in Portland, where the zoo fed Facebook and YouTube a steady diet of videos. The only difference was in the comments section, where her Ohio fans reminded everyone that Nora was, and always would be, their "Buckeye bear." Her love of people was great for social media and for the crowds. But it wasn't normal.

As 2016 turned to 2017 and winter turned to spring, Nora paced less, either because of the medication or because she was settling in now that Tasul was gone. But she was still prone to tantrums. She barked and batted her food bowls across the floor. When frustrated or anxious, she turned her back on her keepers, shunning them. The fits were hard for the keepers to watch, but all animals, not just humans, need to learn to deal with adversity. Her handlers talked constantly about how to help her. Soon after Tasul died, Jen DeGroot and the other keepers started a series of relaxation exercises they called "Zen sessions."

On a rainy day in December, DeGroot prepared for one of the sessions. In the zoo kitchen, an industrial space tucked behind the polar bear enclosure, she loaded a pie tin with lake smelt. Nora was only a year old, but she could read people and their energy. DeGroot took a series of deep breaths to center herself.

A few steps from the zoo kitchen, down a corridor lined with barred doors, DeGroot called out to Nora, who came lumbering up to an opening. DeGroot knelt on a piece of cardboard to insulate her knees from the cold concrete and began to offer fish through a gap in the bars.

No words were spoken, but Nora lay down immediately, head resting on her fluffy paws. DeGroot began feeding Nora the smelt at five-second intervals. To a casual observer, it wouldn't look very Zen, but DeGroot took careful mental notes. Between fish, the cub let out a low growl and shook her head. Her vocalizations were important. If she could wait for the next fish without stressing, it meant she was learning patience. Nora was still in a critical stage of development. What she learned during her first two years would dictate how she interacted with other bears, which would be important for companionship and, someday, breeding. And it would affect her quality of life, which, to her keepers, was most important.

The keepers practiced the Zen sessions with Nora every day. The symptoms of her mental struggles—the pacing and tantrums—improved through winter and early spring, but Nora was still missing something that her keepers couldn't provide: the companionship of another bear.

Before Nora arrived, the zoo had been planning to tear down its polar bear facility and build a new one. Demolition had been postponed twice for Nora's sake. With Tasul gone, the zoo had to weigh its options. With construction looming, it wouldn't make sense to bring in another bear. They'd be shipping Nora out.

Chapter 14

The Last Skin Boat

Gene Agnaboogok's father, Roland, hunted the waters off the Alaskan coast in a boat made from walrus skin, stretched taut over a wooden frame. Today, the skeleton frame of that skin boat, the last one used in the village, sits among the grasses just down the road from Agnaboogok's house, slowly sinking into the permafrost, unused for years.

Hunters use aluminum boats with outboard motors now. The handheld harpoons used by Agnaboogok's ancestors have been replaced by dart guns. Sleds are still prevalent, but they are pulled by two-stroke engines instead of dog teams. The humming of motors bounces off the mountain and is carried above the village by the wind as snowmobiles and all-terrain vehicles ferry the people of Wales around town and out onto the ice to search for prey.

Hunting remains an important part of how the people in the village get by. Food from the two stores in the village can be prohibitively expensive. A ten-pound bag of sugar goes for around $20. A roll of aluminum foil can set you back nearly $14. Gas for the snowmobiles and boats and all-terrain vehicles costs nearly $7 a gallon, sometimes more. Gilbert Oxereok uses complicated calculations to line up his shots when he's hunting. The arithmetic needed to decide whether to spend his money on expensive food from the store or ammunition for hunting provides a much simpler equation. "You take a few cents for a bullet to a couple hundred dollars' worth of meat," he says. "Do the math."

When Agnaboogok runs low on money, sometimes he finds meat outside of the store, past its expiration date but still frozen, which he cooks with packaged seasoning on the hot plate he keeps at home. He sold his shotgun, but the neighbor who bought it sometimes brings him freshly harvested geese. A friend who didn't know how to skin the animal brought him a fresh Arctic hare, which Agnaboogok lamented had been shot with a high-powered rifle, causing blood to clot around the wound and ruining the flavor. "I might fry the liver," he said as he stripped the skin off the hare and removed the legs before getting ready to boil the rest of the meat. "They don't taste right when it's shot up. In the soup it will come out." He's getting older and hopes to stop working soon. When he does, he says he'll probably apply for food stamps to help him get by. He's gotten government assistance for food before, but he rarely renews his application when the benefits lapse. "Too much paperwork," he says.

On paper, Wales looks financially depressed. According to the most recent estimates, roughly 12 percent of the work-eligible population doesn't have a job—twice the unemployment rate for Alaska and nearly three times the national rate prior to 2020—and the annual median family income is just over $31,000. Per capita yearly earnings hover just below $13,000. According to federal guidelines, more than 40 percent of the population of Wales lives in poverty. But statistics can't measure all aspects of quality of life, and unemployment rates don't account for hunters who make a sizable portion of their living off the land and sea.

Plying the waters and ice and land around the cape for game isn't just a means to an end, though it is that. Hunting in Wales, as in the numerous other villages in rural Alaska, is part of the community's identity. It's how Agnaboogok's and Oxereok's grandparents—and their grandparents' grandparents—learned to survive in a climate most would find inhospitable. When one hunter speaks of the whales he's harvested—he's harpooned seven, but three got away—there's a sense of accomplishment in his voice when he says, "I fed my community four times."

Hunters in Wales don't practice all the traditions or observe all the taboos that their ancestors did. Agnaboogok's parents told him that, long ago, after a hunter killed his first polar bear or wolverine, the hunter would isolate himself for a day or longer, avoiding all contact with friends and family. "I couldn't quite understand why," Agnaboogok said. "They had all kinds of beliefs back then." Oxereok still keeps some of the old traditions, but he doesn't like to talk about them.

Like Agnaboogok, Oxereok was taught to never let an animal suffer needlessly and to never hunt more game than was needed. He doesn't go after seals that still have suckling pups. If there is a herd of walrus and his crew has been lucky enough to land one, he'll shoo the rest away. Hunters in Wales do their best to use every part of the animals they kill. The bones, tusks, and skulls are carved or preserved or used in crafts. The meat, the fat, and the internal organs are all consumed. It's part of the respect for the natural world that was ingrained in them from an early age, Oxereok says.

"Respect the animal you hunt. Because it's feeding you and your family. Same way you respect a person, you respect the animals, too. If you don't, it'll come back and bite you."

Hunters in Wales started noticing changes to the climate years ago.

The walrus and seals come past the village sooner than they did when these men hunted as teenagers. Ice that used to support the weight of a hunter and a snowmobile well into spring now flexes and sinks weeks earlier. Agnaboogok used to have to hack more than four feet through ice to hit water; now, in most years, it's more like two feet. Oxereok sees bare ground where he used to see snowbanks. He sees insects he hasn't encountered before and species usually found farther south, in warmer climates. He's noticed an increase in the number of seal pups, which he calls "future food," abandoned on the beach. He read in *National Geographic* that shipping and oil drilling and mining are expanding in the Arctic as the waters, in some places ice-free for months out of the year now, become easier to navigate. He worries that increases in ship traffic and industry will pollute the sea. He's already seen troubling signs. Oxereok pulls seals from the water that are coated in oil and covered in sores, victims of parasites that thrive in warmer temperatures. He can't eat seals like that. "We just slit the belly and let them sink."

As the Arctic climate warms, the extent of sea ice is trending downward. The ice that does form is thinner and breaks off from land earlier, which increases drift. Agnaboogok remembers one hunting trip where he and his crew killed a walrus. They towed it to a low and flat iceberg, the most suitable place on the open water to butcher a creature that likely weighed more than a ton. Cutting up an animal that large is not a speedy endeavor, and in the time it took to chop up the walrus, the iceberg had drifted twenty miles, rocking in the waves the whole time. Agnaboogok was afraid and a long way from home. Everyone made it home safely, but it took them extra time and, more important, extra gas, which isn't cheap.

This phenomenon doesn't just affect hunters butchering walrus. It adds an extra burden for polar bears, too. Using data from radio collars, researchers at the U.S. Geological Survey compared Alaskan bears' movements in the period 1987–98, when sea ice was more stable, and 1999–2013, when the amount of drift had become greater. They found that the bears exerted more energy in the later period of the study to counteract the drift. Energy for polar bears comes from seals, and the bears needed to consume one to three more seals per year to make up the deficit, a 2 to 4 percent increase, while the sea ice they relied on to catch their meals was declining. Food, for people and polar bears, was getting harder to come by.

In 2014, the Inuit Circumpolar Council, a nongovernmental organization that represents 180,000 Native people from Alaska, Canada, Greenland, and the Chukotka region of Russia, published a regionwide assessment of food security around the Bering Strait. They tapped local experts with extensive traditional knowledge and held workshops in more than a half dozen villages, including Wales. Many of the participants were frustrated with the lack of decision-making power they had in managing their traditional prey species. Decisions were being made thousands of miles from the waters of the Bering Strait that directly impacted the people there, with no input from them. Food has value beyond the calories it contains in cultures around the globe, but it has a special importance in communities that rely on hunting for survival. Education plays a prominent role in how Natives secure their food, with some participants in the workshops lamenting that so-called Western schooling was often given higher priority than learning traditional ways of hunting. More time spent in classrooms meant fewer opportunities for children to learn the ways of their elders. Learning civics and history and math was important, but so was learning how to butcher a walrus or how to preserve the meat so it wouldn't spoil in storage.

One of the biggest topics of conversation in the workshops was the changes many had seen in the environment and the direct impact those changes had on food security.

In Gambell, a village just shy of seven hundred on St. Lawrence Island, in the Bering Sea, the winter and spring of 2013 saw a long stretch of unusual winds out of the southeast, along with a spate of wet storms. The weather pushed huge pressure ridges of ice up to the coast, preventing walrus crews from reaching hunting grounds offshore. Between Gambell and Savoonga, a community roughly forty miles to its east, twelve hundred walrus are harvested in an average year on St. Lawrence Island. That year, they got only a third of that. An economic disaster was declared, and a nearby nonprofit donated two thousand pounds of halibut and salmon—about three one-gallon Ziploc bags of fish per family in the region. Two years later, hunting conditions were again poor. Ten thousand pounds of frozen fish were distributed to Savoonga, Gambell, Diomede, and Wales to try to offset the low spring harvest of walrus, but the donated food was just a fraction of what a normal harvest would have brought.

Nearly everyone in the food security workshops said winter was getting shorter and that in many places the ice was melting earlier. Seals and walrus, which need the ice to haul themselves out of the water and raise their young, followed it as it receded north, meaning that hunters had to travel farther to find them. Beaches, usually protected by shorefast ice, were battered by waves, causing erosion and washing out hunting camps. The weather had become less predictable, leaving hunters with smaller windows in which to catch their prey. Where once they saw calm, stable weather for up to two weeks at a time, now they are lucky to have a twelve-hour stretch without a storm rolling in.

Tribal elders in Wales say that children still learn to tell when a fox is rabid, to track a moose, to kill an animal without letting it suffer, and to use every part. Oxereok learned by listening to stories his uncles told, then by earning his way onto a hunting boat. Now he teaches his nephews. It's been that way in Wales for hundreds of years, the older hunters passing knowledge on to the younger generation. But as the skin boat Agnaboogok's father used sinks into the tundra, the older hunters see in the young people of the village a diminishing interest in the ways they were raised with, at least partially because of the changing climate. With shorter winters and volatile weather, opportunities to learn are fewer and farther between.

But there is more than one factor at play. Oxereok worries that traditional knowledge is being lost on the young people as satellite television and social media make inroads into more aspects of everyday life in the village. It's a dynamic common between generations, in urban centers and rural villages alike, but it has more urgency for people in a place like Wales. Oxereok worries that self-reliance, a defining characteristic of those who have populated the cape since it was settled, is fading as more people rely on government assistance to buy food from the store.

"Sooner or later, that government money isn't going to last. Food prices are going to jump, especially with the fires and droughts happening down [in the] states," Oxereok tells the young people in Wales. "When you grow up you're going to be hungry."

Chapter 15

Another Hurdle

One April morning in 2017, Nicole Nicassio-Hiskey watched Nora walk up a ramp toward her den. The young bear seemed playful as usual, but her gait was off.

That's really weird, Nicassio-Hiskey thought to herself. *That's new.*

Nora was favoring her front left leg. It looked as though her elbow was bowing out just a little. Nicassio-Hiskey called the zoo vet, who came by later that day. By the time Mitch Finnegan arrived, Nora's condition had worsened. She couldn't put pressure on the leg, and her elbow jutted out as she walked. Finnegan hoped she was just sore from playing—Nora had developed a habit of jumping into the pool and slamming against the window with her front paws—but he suspected there was something more.

Finnegan knew about Nora's history with metabolic bone disease, so he asked the vets in Columbus for the bear's old X-rays. In mid-May, he sedated Nora and took a set of X-rays himself. While Nora was knocked out, the vets performed a few routine exams. Nora's nails and coat and teeth all looked good. Her heart rate was regular. Her ears and nose were clean and free of any irritation. Her claws were evenly worn and symmetrical. But when Finnegan went to position Nora's left arm for the radiographs, he noticed stiffness in the elbow. Her arm had "a slight angular deformity," he wrote in her chart, "which kept the elbow from lying flat."

Once Finnegan got her arm situated, he took X-rays of all her joints, her skull, her spine, and her abdomen. What he saw was that one of the bones in her forearm was stunted, and because the pieces didn't fit right, her elbow was being forced outward. He sent the X-rays to other vets and orthopedic surgeons around the country. One of them, Erika Crook, a veterinarian at Utah's Hogle Zoo, in Salt Lake City, didn't mince words about Nora's condition.

"All joints are trashed," she wrote in a bulleted list of Nora's ailments. She had "marked dysplasia" in one of her elbows. One of her radial bones—one of the two large bones of the forearm, the same one she'd fractured back in Columbus—was straight, but the head was crooked. The smaller carpal bones in one of her wrists looked like they might have become deformed when she was younger. "She is or will be an arthritic mess, especially because she is such a big animal." The surface where Nora's bones met was "so F-ed up," the vet wrote.

Joints need to fit together with a high degree of precision, especially in an animal as heavy as Nora, who now weighed close to 340 pounds. If the bones don't line up perfectly, like a piston in a cylinder, the joints wear out. Bone growth is governed primarily by growth plates, areas of rapidly dividing cells at either end of long, weight-supporting bones like the radius and the ulna in the forearm. If the growth plates don't get all the nutrients they need during early development, important bones can become misaligned. That was happening to Nora. During those few weeks back in Columbus when she didn't get sufficient vitamins and calcium, Nora's left elbow had become deformed. Her keepers in Ohio thought they had corrected the problem, but it appeared that the longer-lasting impacts had been hidden.

The vet from Utah laid out a couple of treatment options, but each of them came with significant risks. In a dog with similar problems, a veterinary surgeon might perform an ulnar ostectomy, removing a section of the ulna to correct for the disparity in bone length. That surgery worked only in young dogs, though. Nora had yet to pass her second birthday, but she was already too old. Some vets used injections of platelet-rich plasma to alleviate the pain of arthritis. Platelets are the part of the blood responsible for clotting and healing, and the theory was that injecting them near damaged cells might speed along the healing process. In Nora, the procedure might provide some temporary relief, but it would not resolve the underlying problems. It, too, was ruled out.

Meanwhile, Nora's anxiety and aggression intensified. It was hard to tell if she had outgrown the dose of antidepressants or if she was acting out of pain, but Finnegan upped her dose of fluoxetine. He added two daily doses of anti-inflammatory pain medications to the schedule of drugs Nora was taking. The experts also recommended she spend as much time as possible in the water to ease the load on her joints. Nora already loved the water, so that fix was easy, if not entirely satisfying. Her keepers would watch Nora closely and do their best to manage her pain. Other than that, there was little they could do.

Around the same time Nora's keepers first noticed the issues with her joints, the staff at Utah's Hogle Zoo was mourning the death of Rizzo, a gregarious female polar bear. The zoo had gone without a polar bear for nearly a decade before they debuted their Rocky Shores exhibit in 2012. The exhibit was crafted in the style of the Polar Frontier enclosure in Columbus, though it wasn't quite as large. It had a massive pool, with tunnels to swim through, and multiple viewing options both above and below the waterline so zoo-goers could watch Rizzo swim and dive and chomp on her favorite snack, whole watermelons, which she frequently brought right up to the glass to eat in front of curious onlookers. She'd gained a following among the public of Salt Lake City, but her keepers noticed that the nineteen-year-old bear was slowing down. She grew lethargic and began having trouble keeping food down. Blood tests and an ultrasound showed her kidneys were failing. She was moved back to the holding area, out of view of the public, and given fluids, but there wasn't much that could be done. She was euthanized in early April.

That left a modern polar bear exhibit and an experienced keeper staff, with no polar bear to care for. Hogle also had a strong conservation record. Polar bears were part of what the zoo called its "Big Six," six species the zoo was focused on conserving. It had partnered with Polar Bears International on education campaigns and supported the organization as it studied maternal polar bear dens on Alaska's north coast. With the Oregon Zoo preparing to tear down its old exhibit, it was decided that, in the fall, Nora would be moving to Utah. And she wouldn't be alone.

The Association of Zoos and Aquariums had found a companion to join Nora in her new home, a bear from Toledo born just about a month after Nora. The Toledo bear's mom was Nora's grandmother. Nora's new companion, though younger than her, was her aunt. After nearly two years of human care and a brief spell with Tasul, Nora would finally have a polar bear her own age to learn from. The cubs had vastly different upbringings, but that was good. Nora's new companion had been raised by her mother and knew more about being a bear than Nora did.

Nora and the new bear were both about to turn two, the age when cubs in the wild leave their mothers and set out on their own. In a twist that seemed almost ripped from a children's story, Nora's new companion was named Hope.

As Portland's spring rains made way for the heat of summer, Nora appeared to be improving, both physically and mentally. She still had what her keepers called a "bulldog stance," both of her front elbows jutting out to compensate for her off-kilter bones. Finnegan, the Oregon Zoo vet, increased her dose of fluoxetine after an increase in aggression, on top of regularly prescribed increases to account for her growth.

In August, Joanne Randinitis, one of the senior keepers from Utah, traveled to Portland to meet Nora and the keepers from the Oregon Zoo. Under a blistering heat wave with temperatures approaching one hundred degrees, the Hogle PR staff interviewed Joanne on a livestream on Facebook. As Nora chased fish tossed by her keepers in the background, Joanne explained that she was there to get to know Nora so the staff in Utah could provide the best care for her. As she spoke, Nora's legions of fans from Ohio filled the comments. "Nora's a Buckeye!!! We will follow her anywhere!!!"

Behind the scenes, Nora had already begun preparing for the trip. A few weeks earlier, her keepers had attached a travel crate to the network of tunnels and dens, out of view of the public. Every morning, Nicassio-Hiskey and the other keepers coaxed Nora into the crate and gave her her favorite treats to get her comfortable in the enclosure that she would leave Portland in. For the first time in a long time, Nora's future looked bright.

That summer Portland had been laboring under a series of stifling heat waves. In June, a stretch of three ninety-degree days peaked just over the one-hundred-degree mark near the end of the month. July and August saw more of the same, with temperatures often hovering in the eighties and nineties and little precipitation to offer any relief: Portland saw fifty-seven days without rain, the third-longest dry stretch on record. The heat peaked in early August, when thermometers topped out at 105, just a few degrees shy of the all-time record. The month ended as the warmest August on record in Portland, and as Labor Day weekend rolled around, the heat was unrelenting.

Keepers at the zoo kept Nora cool by letting her swim as much as she wanted, giving her frozen fish snacks, and offering up tubs filled with ice cubes, which she rolled and romped in, to the delight of her fans. The people of Portland sought refuge from the heat in public fountains or by driving to the coast. Some headed to the nude beach on Sauvie Island, north of town on the Columbia River. Others headed to the Columbia River Gorge, a four-thousand-foot-deep canyon lined with hiking trails through shady temperate forests and fern groves, many of which lead to picturesque waterfalls and swimming holes.

That's where Liz FitzGerald was headed in early September as she drove to the foot of the Eagle Creek Trail. Her destination was Punch Bowl Falls, a thirty-five-foot cascade that drops into an almost perfectly circular pool, surrounded by basalt cliffs covered in moss that thrives in the mist spraying off the falls. Temperatures crept toward a hundred degrees, and the two-mile trail was packed with sweaty hikers.

FitzGerald had only gone about a mile and a half up the trail when she came upon a half dozen or so teens standing near the edge of a ravine. The National Weather Service had issued a red flag warning for the area, due to increased fire risk, and several miles up the trail from where FitzGerald encountered the group, access was blocked due to a wildfire that was already burning.

FitzGerald watched in shock as one of the teens, a fifteen-year-old boy, lobbed a firework off the cliff. A girl stood by filming with a cellphone. FitzGerald thought she heard them giggle as what looked like a smoke bomb disappeared into the bone-dry trees below.

She kept hiking, but just a few minutes later she looked back and saw smoke billowing up from the way she'd come. She turned back and saw the same group of teens milling about.

"Do you realize you just started a forest fire?" she asked.

"What are we supposed to do about it now?" one of the kids responded nonchalantly.

FitzGerald hustled down the trail, telling two hikers she encountered to turn around and go back. When she got to the trailhead, she flagged down a Forest Service officer and related what had happened. The officer quickly called it in, and just minutes after the initial report, the first helicopter over the fire reported that it had already grown to fifty acres and was cresting the ridge above the trail. Air tankers started dropping water and retardant soon after as panicked hikers streamed down the trail into the parking lot. Faced with steep slopes and a forest filled with tinder-dry trees, there was little that could be done to suppress the fire. One firefighter hiked up to see if the trail was still passable, but the fire was burning on both sides and hot boulders were rolling off the slopes, landing with a sizzle as they plopped into the creek. The trail was cut off by wildfire above and below Punch Bowl Falls, where more than 150 hikers who hadn't gotten out in time were now trapped.

As night fell, it became clear that extracting a group of that size by helicopter would be impossible. A Forest Service worker who had been at the smaller fire farther up the trail hiked down to meet the group, bringing water and supplies to the tired, hungry, nervous hikers. She told them not to worry, that she was in contact with folks who were closely monitoring the fire and they weren't in any immediate danger. They hunkered down, sharing the food they had as they waited for daylight.

Meanwhile, the fire raged. By 2:30 A.M. it had burned down to Interstate 84, a major east–west trucking route and the only road on the Oregon side of the gorge. A half hour later, it was burning in the hills above Cascade Locks, a town of twelve hundred on the banks of the Columbia River, and evacuation orders were issued. As day broke, the hikers trapped on the trail began a long trek out of the forest, taking a circuitous route to a lake. There, they were met by school buses that ferried them back to their cars at the trailhead, where many were reunited with anxious family members.

That second day of what was now being called the Eagle Creek Fire saw only limited spread, given the conditions. There was still a red flag warning in effect, triggered by high temperatures and low humidity, and the fire tripled in size from one thousand to three thousand acres. On Labor Day, the third day after the initial spark, strong winds built from the east and were squeezed through the funnel of the gorge like air through a bellows. Firefighters described the growth of the blaze as "explosive." Embers, picked up by gusts of up to forty-five miles per hour, were starting spot fires more than a mile in front of the main fire line. There was little that could be done from the ground, given the steep terrain. Firefighters continued their assault from the air, dumping water and retardant on new fires as they popped up, but nature was working against them. At 1 A.M., it was still ninety-one degrees with just 24 percent humidity. Around 3 A.M., a twenty-five-acre fire was spotted on Archer Mountain, across the gorge on the Washington side. The Eagle Creek Fire had jumped the Columbia River.

The same winds that fed the flames pushed smoke into Portland, tingeing the skies brown and giving the city an apocalyptic feel. Residents woke up to a thin blanket of what looked like snow on their cars as flakes of ash fell from the sky, carpeting lawns and sidewalks and the concrete floor of Nora's enclosure at the zoo. Air quality plummeted and people were told to avoid outdoor activities. Hardware stores ran low on masks that could block the microscopic particulate matter in the air. The smoke was especially bad for vulnerable groups—homeless people who had no indoor refuge, older folks, and those who suffered from respiratory illness. The Eagle Creek Fire was far from the only one burning in the state, and vast swaths of Oregon suffered under smoky skies for much of the summer. Visits to emergency rooms and urgent care facilities were 86 percent higher than historical trends predicted statewide. On September 5 alone, there were more than 580 asthma-related visits to medical facilities, a 20 percent increase from what was expected for that day. Between August and October, schools across the state were forced to cancel more than 350 outdoor football and soccer games because of poor air quality. The world-renowned Oregon Shakespeare Festival, in the southern part of the state, canceled nine performances, a direct loss of nearly $400,000, and small-town economies suffered as road closures, evacuations, and news stories about the unhealthy air frightened tourists away.

The Eagle Creek Fire would burn 48,000 of Oregon's most beloved acres before it was fully contained, nearly three months after it started. Four homes burned down and four firefighters suffered minor injuries. The fifteen-year-old boy who started the blaze was arrested, thanks to Liz FitzGerald, and pleaded guilty to twelve misdemeanor charges, including reckless burning, criminal mischief, and reckless endangerment. He was sentenced to five years of probation and nearly two thousand hours of community service, and ordered to pay $36 million in restitution, though it is unlikely the full amount will ever be paid.

A week before the fire in the gorge, Hurricane Harvey dumped sixty inches of rain on southeastern Texas, where at least sixty-eight people died as a direct result of the storm. A month later, Hurricane Maria hit Puerto Rico as a Category 4 storm, knocking out power to the entire island. Though the recorded death toll remains in dispute and will likely never be truly known, it is estimated that thousands perished. October saw a series of fires tear through large urban areas north of San Francisco. Dozens were killed, and thousands of homes were destroyed.

The Eagle Creek Fire was far from the biggest or deadliest or most destructive disaster to hit in 2017. But the blaze in the gorge took its own emotional toll on the people of the Pacific Northwest, as told by Jamie Hale, a travel writer for *The Oregonian*:

The gorge isn't just a recreation area, a National Scenic Area, a pretty place to see in the summer. For many of us, the gorge is our temple, our sanctuary, our home. It's a place we go to worship the land, to bask in the awe of nature—where the mightiest river in the west carves through the Cascade Mountains, water pouring toward it in towering falls. . . .

We all know what it feels like to be humbled by a mountain or to fall in love with a sunset. It touches our spirit, our heart and our soul. In the Pacific Northwest, especially, it's a religion—our way of being, wrapped up with the Earth.

We mourn for the gorge. And for all other temples destroyed by our recklessness.

Wildfires and hurricanes are usually referred to as natural disasters. And they are a natural phenomena, the result of atmospheric forces interacting with one another in the physical world. But these types of extreme weather events, for all their destructive power, are not disconnected from the influence of human action. In the case of Harvey, flooding was exacerbated by the urban sprawl that had laid concrete over what was formerly prairie, giving rainwater few places to absorb into the soil. In Puerto Rico, mismanagement of the territory's electric utility and a backlog of delayed repairs and upgrades left the island's power grid vulnerable to outages well before the first winds blew toward the coast. Today's wildfires burn more ferociously at least in part because they were preceded by a century of aggressive suppression that left forests overloaded with fuel. Local planning decisions made what were already sure to be catastrophic events even worse.

But by 2017 scientists had finally begun to understand the role of other human actions on extreme weather. For decades, people like James Hansen had warned that, as greenhouse gas concentrations built in the atmosphere, the climate would warm and extreme weather events would become more common. When people issued those warnings, though, they did so in general terms, always careful to note the difference between climate—the long-term trend of the environment in which we live—and weather, which means day-to-day fluctuations in clouds and wind and precipitation.

But as climate modeling technology advanced, it became possible to tie wider trends in climate to individual weather events, to measure the intensity of storms and droughts and heat waves and look for the fingerprints of climate change. In 2003, a sweltering heat wave blanketed Europe and temperatures topped a hundred degrees for days on end. In many places, that type of heat was unprecedented and few people were prepared. The resulting death toll was catastrophic: Some researchers argued that as many as seventy thousand more people had perished than in a typical year, an outsize proportion of them older. One group of scientists deduced that as a result of greenhouse gas emissions, the type of heat wave that killed so many in 2003 was at least twice as likely to occur as it would have been in a world without warming. In 2016, a team of researchers ran models to figure out how the heat wave might have looked in a world with zero emissions from human activity. In London, they found, climate change increased the probability that the heat wave would be fatal by 20 percent. In Paris, which was far hotter than the United Kingdom, the likelihood was increased by 70 percent. Of the 705 deaths in Paris resulting from the heat wave, more than 500 could be directly attributed to climate change. Researchers applied similar methods to Hurricane Harvey and found that rainfall totals for the storm were as much as 38 percent higher than they would have been in a world that wasn't warming.

Wildfires are not triggered by climate change. Neither are hurricanes, droughts, floods, or heat waves. Most forest fires are started by people—an errant cigarette butt, an unattended campfire, or a smoke bomb thrown by an irresponsible teenager can be all that's needed to set off a conflagration. But the conditions under which a wildfire burns play a significant role in its intensity. In the West, average temperatures in forested areas have gone up by roughly 2.5 degrees Fahrenheit since 1970, sapping moisture from trees, dead vegetation, and soil and leaving forests filled with fuel that is primed to burn. When a spark does land in that dry tinder, fires fueled by climate change burn bigger, hotter, and longer. Using climate models and historical drought indexes, researchers from Columbia University and the University of Idaho found that, between 1984 and 2015, human-caused warming led to an additional ten million acres of scorched wilderness in the West.

* * *

Polar bears can detect a seal from miles away, so Nora must have inhaled the scent of the mountain goats over in the Great Northwest exhibit, the hot fried dough of the elephant ear cart, the orangutans dangling from their fingers in the Red Ape Reserve, and even the wild coyotes roaming the West Hills of Portland. As Nora prepared to make her last appearance at the Oregon Zoo, she would also have smelled the smoke from the Eagle Creek Fire, burning fifty miles away.

She padded gingerly around the yard that day, her limp imperceptible to those who didn't know to look for it. She pressed her face against the bars that led to her swim flume, where a researcher had measured her oxygen use on the zoo's underwater treadmill. On land, she moved awkwardly, her back feet pigeon-toed and forelimbs slightly bowed, but in the water she was majestic. She clambered out of the pool and shook a cylindrical aura of droplets from her thick fur. She blinked in the sunlight fracturing off the turquoise water. She loved to snag the head of a bristle brush and position herself with her rump in the pool. With a swift arc of her neck, she would fling the toy over herself and then execute a back dive after it as crowds of onlookers shrieked.

Over in the zoo's gift shop, Nora's name had disappeared from the polar bear merchandise as her time in Portland neared an end. All that remained were a few polar bear baby bibs, a smattering of refrigerator magnets, and the last of the shiny ceramic figurines relegated to a remote corner of the store. It was a mild Sunday, a respite from what had been a scorching summer, and the winds that had inundated Portland with smoke had shifted, clearing the skies. That morning, the crowds streamed into the Pacific Shores exhibit, hoping for one last glimpse of Portland's famous polar bear cub. Nora obliged her fans, plunging into the pool and romping through the yard as her keepers tossed down cardboard boxes wrapped like going-away presents, PDX TO SLC SEPT. 2017 printed on the sides.

"Nora is going to a new home in Utah to be with another bear," a volunteer explained to the crowd. He made no mention of the extent of her emotional troubles or the ailments that had deformed her bones as she splayed out, napping on a mound of manufactured snow. "She may not understand the importance of this day like we do."

In the back of the viewing area, spinning placards showed visitors "10 things you can do to help save polar bears," changes people could make in their daily lives to reduce their carbon footprint. Maybe those changes would make a difference at scale, if everyone decided they would make a concerted effort to drive less or remember to turn off the lights, but, faced with the prospect of hurricanes like Harvey and Maria and fires like the one burning in the gorge and the changes to the ecosystem that were threatening the food sources of the people in Wales, the suggestions seemed so small as to be almost meaningless.

A toddler fiddled with the first sign as his father sat on a bench posting a video of Nora online. A few feet away, a globe showed the Arctic sea ice in stark relief. "That's where it was in 2005," another father said to his son, pointing to a line on the map that circles the top of the globe. "And that's where it was in 2016," he said, pointing to a much smaller circle.

"It's all shrinking," the boy replied.

Chapter 16

On the Edge of a Warming World

Northeast of Wales, the village of Shishmaref is disappearing.

Shishmaref is about seventy miles up the coast from Wales, and it's larger, home to roughly six hundred people. But it has much in common with its neighbor to the south. There's an airstrip and dirt roads that wind past prefabricated houses and around the community's two stores, the church, the school, and a bingo hall. The people of Shishmaref live similarly to those in Wales, too. There's work to be had, but many prefer to make their living off the land and sea, hunting for game on the sea ice when it's there, fishing or gathering greens and berries when the ice breaks up.

But where Wales is planted firmly on the mainland, Shishmaref sits precariously on a sandy barrier island called Sarichef, one of dozens that line the coast of the Seward Peninsula. Sarichef is two and a half miles long and a quarter mile wide. A sizable saltwater lagoon washes against it on one side, the open ocean on the other. The highest point on the island, composed of sand dunes, stands roughly twenty-five feet above the sea. It's been that way for hundreds of years, probably longer. Until recently.

Erosion has been a problem on the island since at least 1950, when residents began filling drums with sand and putting them on the beach to buffer the effects of the pounding surf. In good years, the village would lose three to five feet of land as the sea steadily ate away at the island. In bad years, like 1973, they lost much more. In September and November of that year, a pair of storms blew off the coast of Siberia and strengthened over the Chukchi Sea. The second of the two storms packed winds of up to eighty miles per hour and produced waves up to fifteen feet high in the Bering Strait. Both storms hit at or near the monthly maximum tide, exacerbating flooding and damage to the coast. Over the course of just a few weeks, Shishmaref lost at least thirty feet of shoreline. The village voted to relocate, and fifty thousand sandbags were filled and placed on the seaward beach as a stopgap measure. But the proposed relocation site had thick permafrost below the soil, a significant hurdle to construction, and as subsequent years brought gentler storms and less erosion, the effort lost momentum.

Over the following decades, the residents of Shishmaref tried a number of mitigation schemes to protect their homes from the encroaching ocean. In the early eighties, they installed gabions, large wire cages filled with rocks, at the toe of a crumbling bluff to protect the homes that sat atop it. A few years later, they covered nearly two thousand feet of the beach in concrete blocks. In the early nineties, more gabions were added. None of the strategies provided more than short-term relief, and in some cases the fixes caused more significant erosion in unprotected areas.

Then the storms came back. A strong weather system in the fall of 1997 took another thirty feet off the island's north shore and Shishmaref was declared a disaster area. More than a dozen homes and a National Guard armory had to be moved across the island, plopped onto skids, and dragged to more stable ground.

An analysis by the state the next year found that as many as twenty-two more homes were at risk of imminent loss from the accelerating erosion. Powerful storms battered the island again in 2000, 2001, and 2002. By then, sea ice had begun to form later and break up earlier than it had in the seventies, meaning the island weathered some of those storms without any buffer against the crashing waves. At the same time, the permafrost on which the community had been built was thawing, softening the land and adding to the erosive power of the ocean. In 2002, the village again voted to relocate, but, as had happened in 1973, potential relocation sites proved unsuitable, and funding for the move—estimated to cost almost $180 million, or roughly $300,000 per village resident—never materialized. Instead, officials funneled $27 million into more coastal protection projects between 2005 and 2009, which were expected to provide a safeguard for the people of Shishmaref through the 2020s. They would, at least for the short term, hunker down and brave the elements.

Like the polar bear, the people of Shishmaref became poster children for climate change. In 2005 alone, more than a dozen media outlets visited the remote community, including international film crews from Germany, France, and Sweden. Correspondents from *Time* and *National Geographic* flew in, too. Future Pulitzer Prize winner Elizabeth Kolbert wrote about the village in the first chapter of her 2006 book, *Field Notes from a Catastrophe*. News sites ran images of homes tipped off their foundations, mid-tumble down the bluff: the effects of climate change happening in real time, not as some ambiguous theory that future generations would have to contend with.

Shishmaref is not alone in the peril it faces. Up the coast in

Kivalina, the closest village to Karyn Rode's former base camp, the sea is eating away at a barrier-island community that is home to around four hundred people. Like their counterparts in Shishmaref, residents of the village voted to relocate, but they have yet to find a suitable location. Near the mouth of the Ninglick River, on Alaska's southwest coast, the village of New-tok is in the process of moving because of melting permafrost and erosion. All told, more than two hundred of the state's Native villages are facing increased land loss and flooding, according to an analysis by the Government Accountability Office. At least thirty-one of those are facing what the government classifies as "imminent threats," and twelve, including Shishmaref, Kivalina, and Newtok, either had voted to move or were considering relocation.

Svante Arrhenius, the nineteenth-century Swedish scientist who was one of the first to warn of the warming effects of greenhouse gases, hypothesized as far back as 1896 that the poles would heat up faster than the rest of the planet. He didn't get all the details right, but the main thrust of his theory has been borne out. Temperatures in the Arctic have gone up at twice the rate as those of the rest of the globe.

Arctic amplification, as it's referred to in scientific literature, happens for a variety of reasons. As the points on the earth farthest from the sun, the poles get sunlight at a greater angle than anywhere else on earth and are naturally the coldest parts of the planet. These low temperatures have, for as long as we've had a stable climate suitable for human habitability, brought snow cover and ice. Those white surfaces reflect sunlight and warmth, keeping temperatures low. As greenhouse gases have trapped more heat, snow cover and sea ice have decreased, leaving exposed areas of tundra and open ocean. Like a dark T-shirt on a hot day, these dark areas absorb more heat, and that heat melts more snow and ice, leading to more areas of open ocean and snow-free tundra that in turn lead to even more heat absorption. In technical terms, the loss of reflectivity decreases the earth's albedo, the ratio of outgoing radiation reflected off an object to the incoming radiation that it absorbs. It's a process that, given a little upward temperature nudge by human activity, creates its own feedback loop.

The decrease in albedo is thought to be the primary driver of Arctic amplification, but there are other factors, too, and not all of them originate in the Arctic. The tropical regions of the earth, for example, are home to frequent thunderstorms. Like their more powerful cousins, hurricanes, thunderstorms are powered by convection, the process by which warm air rises and becomes unstable. Once that warm air is sucked into the upper atmosphere, it is carried by steady wind patterns toward the poles. Because thunderstorms are common in the tropics, the transport of heat from the equator to the poles is reliably consistent. It keeps equatorial regions from overheating, but it also adds to Arctic amplification.

Changing temperatures in the Arctic are likely to have a cascading effect on the ecosystem. At the base of the Arctic food web are some of the ocean's smallest creatures, minuscule crustaceans called zooplankton. They provide nutrition for a wide range of species, from forage fish like the Arctic cod to seabirds to massive marine mammals like bowhead whales. The creatures that eat those creatures, from people to polar bears, ultimately rely on the zooplankton for their meals.

Zooplankton come in a wide variety, though. While they are tiny, some, like *Calanus glacialis*, which measures less than a quarter inch at its largest, are particularly rich in fat and other essential nutrients. But *Calanus glacialis* feeds primarily on algae that form on the underside of sea ice and are released into the water during the spring melt. If the melt comes too early, the copepods like *Calanus glacialis* could miss their best window for bulking up. Researchers have found that cold years with lots of long-lasting ice tend to see a more diverse population of zooplankton swimming in Arctic waters, while warm years with less ice have a more monolithic population and fewer *Calanus glacialis*.

If greenhouse gas emissions continue on their current path, the Arctic could look very different by midcentury. Some animals will be able to adapt—warmer years could bring different species of zooplankton that can fill any nutritional gaps left by their colder-water cousins, but they might spawn in different locations or at different times of year, altering the migration patterns of the animals that feed on them. Others, especially those dependent on sea ice, face a bleak future. Polar bears differ across their nineteen subpopulations, but all of them need a platform of sea ice from which to hunt seals, the only prey with a high enough calorie count to sustain them. With less ice, bears have to travel farther for meals, expending more energy, and the amount of time they spend on land waiting for ice to form will lengthen. And because a female polar bear's body will produce a cub only if she has sufficient fat reserves stored for the long winter she'll spend in the den, fewer days on the ice means fewer seals means fewer cubs.

The most recent research, published in 2020, found that even with moderate mitigation of emissions, some subpopulations will begin to experience reproductive failure and localized extinction by 2040. By the end of the century, if emissions continue unchecked, polar bears will vanish from the entirety of their range, with the exception of the northernmost islands in the Canadian Arctic.

But a melting north has far-reaching implications for life well south of the Arctic Circle, too.

* * *

The world's climates can be roughly divided into zones. Some of the warmest climates, the tropics, are around the earth's equator. Heading north, you hit the temperate zone, located in the mid-latitudes, home to fewer extreme climates. Above that is the Arctic, which plays an outsize role in maintaining stability in the mid-latitudes.

In the Northern Hemisphere, the temperature difference between the mid-latitudes and the poles drives the polar front jet stream—a high-altitude river of supercharged winds that circle the earth from west to east around the same latitude as the border between the United States and Canada. A strong jet stream blows fast and straight, but a weak one can wobble, allowing cold air to dip to lower latitudes and warmer air to travel north. A weakened jet stream is also more likely to get stuck in place, causing extreme weather events to last longer. As the north has warmed and the temperature difference between the Arctic and the temperate regions has diminished, scientists have found evidence that the jet stream is weakening, and some studies have shown links between the slowing jet stream and events like the 2003 European heat wave that killed thousands, as well as an unprecedented melt-off of the Greenland ice sheet in 2015.

As the glaciers that fill high-altitude valleys and the ice sheets that cover Greenland and Antarctica start to melt, every drop of water that is added to the ocean raises sea levels around the globe. Even under the most optimistic scenarios, sea-level rise still poses a grave threat to people living on the world's shorelines and in low-lying island nations. Residents of places like the Maldives and Tuvalu, island chains that barely rise above the seas in which they sit, are likely to see their farmlands swamped and their beaches washed away. Without the implementation of costly adaptation plans, flooding damage worldwide is expected to increase by two to three orders of magnitude by 2100. Large swaths of East Asia—including huge population centers like Shanghai, Bangkok, and Ho Chi Minh City—will likely see frequent flooding by midcentury. A 2019 study estimated that as many as three hundred million people currently live in areas that will be subject to chronic flooding within thirty years. By 2100, areas that are now home to two hundred million people will fall permanently below the high-tide line. Seawalls and flood protection may be possible in large urban areas, so long as the government is willing to pay for it. For smaller, rural places, finding that kind of investment will prove challenging, and communities may be left with few options other than to relocate.

Esau Sinnok was born in Shishmaref in 1997, the same year the powerful storms returned and again ate away at the shoreline in large chunks. He lived with his grandparents, but he was raised by everyone in the village—uncles and aunts and cousins and elders who told him stories of what the island was like when they were young. Like Gene Agnaboogok and Gilbert Oxereok, in Wales, Sinnok was out in hunting boats when he was still a toddler, and he remembers his uncle Norman Kokeok taking him to hunt birds and gather eggs when he was just four or five years old. Around the age of seven, he got his first seal and began to learn the yearly cycles of hunting in the Arctic.

In the spring and fall, they hunted ducks and geese on land, seals and walrus on the Chukchi Sea. They gathered berries when they ripened and hunted caribou year-round. Sinnok learned the importance of hunting, not just to feed his immediate family but to help provide for those who didn't have access to a boat or a snowmobile or who were too old to venture out on the ice. He learned much of that from his uncle Norman, who moved to Fairbanks for a few years but returned to Shishmaref when his parents got too old to hunt so he could help provide for them.

When the elders told Sinnok stories about what Shishmaref was like when they were young, it sounded like they were describing a different world. When he was a kid, Sinnok's grandpa told him, the ice stretched twenty or thirty miles out into the sea. Sinnok doesn't remember there ever being more than a mile or two of sea ice.

In 2006, a nasty storm sent waves crashing over the roof of Sinnok's grandparents' home, a blue house that sits on a bluff overlooking the Chukchi Sea. To Sinnok, it felt like the building was going to collapse. The house next door did, tipping off the sandy cliff and coming to rest at an unnatural angle, half on solid ground, half on the beach. His grandparents' home was one of the many that had to be moved across the island.

The following year, Sinnok's uncle Norman was coming back from a hunting trip in early June. It was summer, but the ice on Shishmaref Inlet, on the leeward side of the island, was usually solid that time of year. As he motored toward home on his snowmobile, the weight of the machine proved too much. It crashed through the ice. It was early morning, around 5:15 A.M., but villagers raced out to help him, performing CPR and trying to revive him. A rescue crew was dispatched from Nome, but it was too late. He was pronounced dead just before 9 A.M. Today, a white cross that marks his grave in Shishmaref bears the words BELOVED SON, BROTHER, UNCLE, FATHER.

Sinnok was barely a preteen, but his uncle's passing and the relocation of his grandparents' home were galvanizing events for him. His ancestors wouldn't have faced the problems his village is facing now. Historically, most coastal Alaskan villages were seminomadic, their hunters moving from winter camps to summer camps, following the migration patterns of the game they hunted and the blooming of the greens and berries they collected. When the federal government began building schools and mandating that Native children attend them, the people living in places like Shishmaref and Kivalina became tethered to permanent infrastructure. Barges needed places to land with construction materials, so the sites for these new buildings were chosen for their accessibility. Once the schools were built, other permanent fixtures followed. Clinics and hospitals. Power plants and community halls. Sewage lagoons and airports. The new amenities came with benefits, to be sure, but they effectively eliminated the villagers' ability to adapt to changing conditions like erosion.

The Shishmaref School, the biggest building in town, clad in gray and green paint, went up in 1977 and now hosts roughly two hundred students, from prekindergarten to seniors in high school. That's where Sinnok began to learn about climate change, about how the emissions coming from smokestacks and tailpipes in faraway places were contributing to the late freeze-up and early thaw of sea ice around Shishmaref. How, with less ice, the brutal storms took a heavier toll on the beaches of his home. How the safety of solid ice, relied upon for centuries, was no longer a given because the Arctic was warming twice as fast as the rest of the planet.

Sinnok decided he needed to act. When he was twelve, he won second place in the state science fair for a project on climate change. In 2015 he became an Arctic Ambassador through a partnership with the Interior and State departments and a local nonprofit. In December of that year, just after his eighteenth birthday, he went to France for the United Nations Climate Change Conference. There he met Itinterunga Rae Bainteiti, from Kiribati, an island nation facing the same threats from rising seas as the Maldives and Tuvalu. The atolls of Kiribati straddle the equator in the central Pacific and have an average high temperature of around ninety degrees. Though they grew up in very different places, Sinnok and Bainteiti share a common threat. "If no one is serious about climate change, we may be underwater as well," Bainteiti told Sinnok.

At the conference in France, world leaders drafted what came to be known as the Paris Agreement. The successor to the Kyoto Protocol, the accord codified a new and urgent goal for the international community: limiting warming to no more than two degrees Celsius by the end of the century. Some amount of warming was already guaranteed, but if we could manage to reduce greenhouse gas emissions enough to meet that threshold, we could avert some of the worst outcomes of climate change. Like previous international treaties, though, the Paris Agreement was nonbinding and allowed individual countries to determine how and to what extent they would reduce emissions. President Obama hailed the agreement as an important step forward, even while acknowledging the accord's shortcomings. "Even if we meet every target embodied in the agreement, we'll only get to part of where we need to go," he said in a later speech.

Even with the adoption of the Paris Agreement, Sinnok was restless. It didn't seem like enough. His uncle Norman and the rest of his community in Shishmaref were still on his mind when, in October of 2017, he joined with fifteen other young people from Alaska and sued the state. At the heart of the lawsuit was a claim that Alaska had violated his constitutional rights by implementing policies that authorized, and in some cases incentivized, the continued extraction of fossil fuels. The suit also cited the Public Trust Doctrine, a legal concept dating back to the Byzantine Empire, which states that certain natural resources are a public good and should be protected for the use of everyone. Named as defendants in the suit were the governor, the commissioner of the Alaska Department of Environmental Conservation, and the Alaska Energy Authority, among other governmental entities.

Sinnok and his fellow plaintiffs weren't the first people to sue over the threats they faced from climate change. In 2008, the villagers of Kivalina—whose barrier-island home is eroding away up the coast from Shishmaref—sued Exxon, along with a host of other oil, gas, and coal companies. The suit alleged that climate change was severely impacting their way of life and sought monetary damages from the companies. That case was dismissed when the court ruled, in essence, that the villagers needed to take their problems to their elected officials to find a legislative fix. The Kivalina case was appealed, then dismissed again. In 2013, the U.S. Supreme Court declined to give the people of Kivalina a hearing, ending the case for good.

But the court battles over climate change were far from over. In 2015, Kelsey Juliana, a teenager from Nora's adopted home state of Oregon, joined with twenty other young people, ranging in age from eight to nineteen, to file suit against the federal government. Among the plaintiffs was Sophie Kivlehan, granddaughter of James Hansen, the climate scientist who issued the grave warning to Congress in 1988 about what was to come. Hansen himself also signed on to the suit, representing "future generations" in court filings. The federal case went through its own series of dismissals and appeals until, in 2019, it went before a three-judge panel on the Ninth Circuit Court of Appeals. In a 2–1 decision, the court conceded that "the federal government has long promoted fossil fuel use despite knowing that it can cause catastrophic climate change," but dismissed the case anyway, because, as in the Kivalina case, remedies to climate change were beyond the power of the courts.

The lone dissenter, Judge Josephine Stanton, excoriated the decision. "In these proceedings, the government accepts as fact that the United States has reached a tipping point crying out for a concerted response—yet presses ahead toward calamity. It is as if an asteroid were barreling toward Earth and the government decided to shut down our only defenses," Stanton wrote. "Seeking to quash this suit, the government bluntly insists that it has the absolute and unreviewable power to destroy the Nation."

Lawyers for Juliana and the other young people vowed to appeal the case to the full Ninth Circuit Court.

There are other lawsuits, too, each of them taking a slightly different legal approach to the issue. But the fundamental question underlying almost all of them is similar: Do our rights to life and liberty and the pursuit of happiness have enshrined within them a right to an environment that makes those things possible? Does a government have a duty to protect its citizens from climate change by regulating the industries that are causing it? What obligation does a society have to protect its most vulnerable from harm?

In 2016, with their options seemingly exhausted, the people of Shishmaref called another vote on whether they should relocate. For the third time since the seventies, the villagers voted yes. The margin was slim, with many of the older residents voting to stay. Sinnok respected their decision: Shishmaref was the only home many of them had known. But as a young person, Sinnok was thinking about the future when he cast his vote to relocate his village to one of two sites a few miles away on the mainland. "It's the future generations, the next seven generations that I want to live in a community called Shishmaref," he said in an interview about the vote. "I want them to grow up the way that I grew up, a traditional subsistence lifestyle."

Sinnok left Shishmaref for college when he was eighteen and

lives in Anchorage now. With a degree in Alaska Native studies, he's planning to move to Juneau, the state capital, to pursue a career in politics. He wants to run a campaign or work as a legislative aide to advocate for the causes that give him purpose: Native rights, justice for missing and murdered Indigenous women, feminism, and, of course, climate change.

"I have family and friends and cousins back at home who are growing up like I did, running around the island they also call home, just seeing it erode into the sea," he said. "We have to do something now."

Sinnok wants to be able to help the people who still call Shishmaref home, someday as their state representative or senator. Someday after that, maybe as governor of Alaska.

Chapter 17

Home, for Now

Utah's Hogle Zoo sits in a neighborhood called East Bench, at the foot of the Wasatch Mountains, which tower over the eastern edge of Salt Lake City. Nora arrived at her new home on a cool day in mid-September 2017. In the weeks before she arrived, the staff at Hogle had drained the 165,000-gallon pool and power-washed the walls, making sure the tunnels and faux-rock perches were clean. They added a shade structure and an extra den so that both Hope and Nora could each have their own place to seek refuge. The zoo purchased training modules, small cages that could be attached to the gates of the dens behind the scenes, so the bears could learn to present their paws, the same way Tasul did when she learned to voluntarily give blood in Portland. Staff members had sat through meeting after meeting as they prepared for the bears' arrival, going over the two bears' diets and conferring with their counterparts in Portland and Toledo.

In the time between Nora's last day on exhibit in Portland and her arrival in Utah, the staff from the Oregon Zoo revealed her bone issues to the public. In a nine-minute video, Amy Cutting, the curator from Portland, explained the issues Nora had as a cub, how she had failed to metabolize certain nutrients in her formula. "She was lucky to be alive," Cutting said as she told

Nora's fans about the difficulties of hand-raising a polar bear. It was the first public acknowledgment of just how dire the cub's condition had been in those first few months of her life and highlighted the hard work the Nora Moms had done to diagnose her ailment and treat it quickly. Cutting talked about the tantrums Nora had thrown when she first arrived in the Pacific Northwest and how the keepers had used Zen sessions and anti-anxiety meds to calm her nerves. She discussed the trouble Nora had getting used to Tasul's presence and described the older bear's last days. She stressed the good Nora could do as an ambassador for her species and how important it was for her to learn from another bear. Nora's issues would be ongoing, she said, but she ended the video on an optimistic note. "Those first critical weeks will affect her joints and her bones as she gets older," Cutting said. "But she's already defied the odds. Nora is a survivor."

Not everyone who watched the video bought into the hopeful theme. "This story is such propaganda," one commenter wrote on Facebook. "There is nothing good about an animal forced to live in a cement cage when it is designed to wander for hundreds of miles with offspring. . . . This is a crass effort to convince the public to buy tickets and view inprisoned [sic] animals for 'entertainment.' A tragedy to watch."

Others were less critical. "Very touching story, so glad she made it!" another commenter wrote. "I hope she lives a long happy life, even in a zoo!"

Nora's long-term prognosis remained an open question. Veterinarians wouldn't know the full extent of the damage to her bones until she finished growing in a couple of years. There was a small possibility that Nora's adult weight would stress her joints and cause so much pain that keepers would have to make what's euphemistically referred to as a quality-of-life decision. If the bear appeared to be suffering and there was no way to alleviate her pain, she would have to be put down. The chances of that were remote, though. Nora was small for her age—Cutting sometimes referred to her as "peanut"—so her body weight wouldn't cause as much stress on her joints as would have been the case for a heavier bear. She would likely deal with some discomfort and a slight limp for the rest of her life, but her keepers were optimistic about her physical well-being. As for her mental health, her keepers were confident that once she got comfortable with her new companion, she wouldn't need anxiety medication anymore.

Zookeeper Kaleigh Jablonski had worked with the Utah's Hogle Zoo's previous polar bear, Rizzo, for almost three years. But the Rocky Shores exhibit, an $18 million project completed in 2012, had been vacant since Rizzo died in April, and Jablonski had imagined the enclosure would stay empty for a while. When she heard that two cubs were coming, she was overjoyed. She'd loved animals ever since she saw a marine mammal show when she was six years old. She'd worked with giraffes, Cape buffalo, and seals before moving to the zoo in Salt Lake City, where she spent most of her time with bears—Rizzo, before she died, and the three grizzlies that live in an adjoining exhibit.

Jablonski knew that Nora and Hope had fans who would be watching to see that everything went smoothly. She'd seen the comments on Facebook and YouTube. She knew how much angst the meetings with Tasul had caused for Nora, and the anxiety that had set in when her keepers left her in Oregon. When the young bear arrived, Jablonski felt the pressure of it all. Then she saw her. Nora walked out of her travel crate and into the small yard where she would spend time alone in quarantine for her first few weeks in Utah. Floodlights illuminated the gravel as Nora cut straight to the pool.

"Hey, baby girl! You're here!" Jablonski squealed. She couldn't help it. Even the fiercest animals at the zoo got the baby talk.

Two keepers from Oregon had accompanied Nora to Salt Lake City. Now they stood to the side.

On that first day in her new home, once Nora had taken stock of her surroundings, she walked past her Oregon keepers and went straight toward Jablonski. Her new keeper crouched on the other side of the metal fence as Nora approached, until the two stood nose-to-nose, faces inches apart, for what felt like forever.

Tears welled up in Jablonski's eyes.

Over the next few weeks, Jablonski and Joanne Randinitis, the keeper who had gone to Oregon to meet Nora a couple months before, did everything they could to acquaint the two bears with each other. Nora and Hope were still in separate holding areas, but the keepers had been passing toys between them so they could get used to each other's scents. By early October, it was time for them to meet face-to-face. Nicole Nicassio-Hiskey, the keeper from Oregon who knew Nora as well as anyone, flew in to watch the introduction.

Around 8:15 A.M., well before the zoo opened, the keepers let Nora into the public-facing enclosure. The holding areas in the back offered nearly as much space, but the yard was big and, more important, open. There were fewer places where either bear could get cornered. The keepers had fed Nora and Hope earlier that day to make sure they wouldn't get territorial over food. Hope outweighed Nora by 180 pounds, but they were the same age, and the keepers were optimistic that they wouldn't see a repeat of what had happened when Nora met Tasul. Just to be safe, however, they had positioned themselves around the enclosure. Some were on the roof, ready to intervene with noisemakers and bags of treats to throw down if the introduction didn't go well.

Others watched from ground level, behind the windows where zoo-goers usually lined the glass.

The keepers opened another door and let Hope into the yard. Randinitis, Jablonski, and Nicassio-Hiskey watched and waited.

It didn't take long for the bears to realize they weren't alone. Soon after Hope walked in, the chase started. But this time Nora wasn't the one retreating. It had been almost exactly a year since she had met Tasul, and in that time her confidence had grown. Randinitis had assumed Hope would be the braver bear, but it was Nora who strode right up to her much larger counterpart. Hope turned and ran and, after a short pursuit, jumped into the pool. She clambered out and the roles reversed, Hope trotting after Nora. For Nicassio-Hiskey, the most encouraging part was seeing Nora walk toward Hope instead of her keepers. Given the attachment issues Nora had dealt with in Portland, she had been expecting the opposite.

After a few more chases and some minor shenanigans—each of the bears snuck up on the other at some point during the first few hours—the bears settled into their respective corners of the exhibit. It was clear to the keepers that Nora was unfamiliar with the cues Hope was giving her and that she was wary. "You can see how stressed out Nora is," Randinitis said as the young bear lay in the sun, her back to a wall, periodically lifting her head to sniff and look around, keeping a cautious eye on Hope.

From the moment they met, the keepers were impressed with how Hope reacted to Nora. She had a sizable weight advantage, but when the two interacted, Hope's behavior was submissive. She was never aggressive with her smaller companion and never intruded into the personal space Nora clearly needed. At one point she even grabbed a toy in her jaws, trotted toward Nora, and dropped it at her feet, almost like a peace offering. Nora seemed confused, still unable to read the body language of another bear. For the first week or so, they tolerated each other but kept a healthy distance. Nora spent much of her time lying in the corner, sometimes seeming to hide behind a pole in the enclosure.

The zoo posted videos from the Rocky Shores exhibit. Nora's fans from Portland and Columbus were met with images of their favorite bear, usually so rambunctious, sitting sedately in a corner. "Poor Nora isn't happy," one viewer wrote. "This really makes me sad. Nora isn't being her playful happy self," wrote another. "She needs to come home to Ohio." Soon after the video was posted, the zoo added a note at the top of the post.

Nora fans—please note—Nora is doing great! She is playful and she is adjusting. Yes, there are times she likes to keep her eye on Hope but that's perfectly normal and they're getting closer everyday. Nora is just like any of us in that we're not all continually playing every second of the day. She's doing great—no need to worry.

In another video update, Randinitis spoke about a recent trip she'd taken to Churchill, Manitoba, on the west coast of Hudson Bay. The town of nine hundred is known for polar bear tourism, and numerous companies run tundra tours when the bears congregate there in the fall waiting for the ice to form. Randinitis had gone there with keepers from other zoos so they could learn to better connect the experiences of animals like Nora and Hope with the lives of wild bears in northern Canada. Randinitis was broadcasting live, and as soon as she mentioned climate change, a familiar phrase popped up in the comments.

"Climate change . . . #fakenews."

* * *

The disinformation campaign waged against climate science has been fraught from the start, and that's kind of the point. Even the names given to certain phenomena that fall under the umbrella of climate change—even that term itself—have been weaponized and warped to confuse and conflate.

For example, there's a long-running myth, promulgated by President Trump, among others, that at some point scientists and environmental advocates, unable to prove that the globe was warming, jettisoned global warming and opted for the vaguer "climate change."

In reality, climate researchers have used both terms for more than half a century, because they describe different things. Global warming is the steady upward march of temperatures caused by the release of greenhouse gases. Climate change is just what it sounds like: changes to climatic conditions—including increased rain from hurricanes in Houston, heat waves in the Pacific Northwest, and the temperature changes that melt sea ice around Shishmaref and Wales—caused by that same buildup of gases.

Whether you call them climate deniers or climate skeptics or what's becoming a popular term in scientific circles, climate contrarians, those who want to distract from or delay action on climate change share similar strategies and goals. They seek to create doubt in an effort to impede action on climate change, to transform what should be a scientific debate about how bad the problem is to a political debate about whether the problem even exists. If you measure it by how much progress has been made in curtailing emissions, the disinformation campaign has been wildly effective.

One of the foremost researchers who has studied these types

of attacks on science is Naomi Oreskes, a world-renowned geologist, historian of environmental science, and professor at Harvard University. In the 2010 book *Merchants of Doubt,* which she co-authored with Erik Conway, Oreskes compares the strategies of climate deniers to the campaign waged by tobacco companies to convince the public that their product was safe despite overwhelming scientific evidence to the contrary.

Casting doubt on the consensus is one of the primary ways that action is delayed. So Oreskes wanted to establish what the consensus on climate change was. By the early 2000s, every major scientific body in the United States that has any bearing on climate change—the National Academy of Sciences, the American Meteorological Society, the American Geophysical Union, and the American Association for the Advancement of Science—had agreed that the evidence for human causation of climate change was compelling. But Oreskes knew that even respected research institutions could sometimes downplay internal dissent over research. So in 2004 she provided the first hard numbers on the exact level of agreement within the scientific community. Among the more than nine hundred peer-reviewed papers on climate change she reviewed, Oreskes found precisely zero that attributed climate change to anything other than human action. The consensus, at least among the papers she examined, was 100 percent.

Even with the results coming down so starkly on the side of consensus, Oreskes left room for some healthy scientific skepticism. "The scientific consensus might, of course, be wrong," she wrote. "If the history of science teaches anything, it is humility, and no one can be faulted for failing to act on what is not known. But our grandchildren will surely blame us if they find that we understood the reality of anthropogenic climate change and failed to do anything about it."

In the face of skepticism, science requires that results be replicable. A 2009 survey found a 97 percent consensus that climate change was occurring as a result of fossil fuel emissions. The next year, another study with a larger sample size came up with the same figure. Over the next several years, four more papers all found the scientific consensus to be between 91 and 97 percent. There is disagreement about how to attribute certain events to climate change, how to most quickly curtail emissions, and which adaptive strategies to implement. But what is certain, for at least nine out of ten climate scientists, is that the planet is warming, and it is doing so because of our actions.

The fossil fuel industry and their proxies in denier groups like the Global Climate Coalition have used the imagined disagreement in the scientific community as a public relations talking point for years. "Emphasize the uncertainty in scientific conclusions" and "urge a balanced scientific approach," reads an internal memo from Exxon in 1988. Victory, according to a 1998 American Petroleum Institute memo, would be achieved when "media coverage reflects balance on climate science and recognition of the validity of viewpoints that challenge the current 'conventional wisdom.'" An analysis of news programs on the major networks and CNN between 1994 and 2005 showed that 70 percent of reports presented viewpoints contrary to the scientific consensus, for the sake of providing "balanced" coverage. Influencers and politicians took notice. From 2007 to 2010, the idea that there was no consensus was the most popular denial argument put forward by the writers of conservative op-eds. In a 2017 interview, President Donald Trump's ambassador to Canada said that there were "scientists on both sides that are accurate," which is not only incorrect in context but logically contradictory on its face.

Contradiction is common among deniers. Cold weather is

proof that the climate isn't warming, they argue, but extreme weather on the other end of the thermometer—heat waves and droughts—doesn't prove anything. They point to the fact that the climate has always changed, which is true, but refuse to acknowledge the rate at which it's currently changing, that temperatures are predicted to rise twenty times faster over the next hundred years than they did during previous periods of warming. They cherry-pick data, choosing specific statistics that support a contrarian opinion, when all the data taken together points to a different conclusion. They chalk it all up to a vast conspiracy, a Chinese hoax, or a socialist plot by the academic elite of the world, a cabal of evil professors intent on redistributing global wealth. They use half-truths and innuendo to air their contrarian views on a host of blogs with names like *Junk Science, Climate Depot,* and *Watts Up with That,* which claims to be "the world's most viewed site on global warming and climate change."

On the internet, there's a niche blog for everything. The consensus within the scientific community that the climate is changing because of human actions is overwhelming, so polar bears often stand in as a proxy for that change, a focal point for those looking to propagate doubt. The internet's polar-bear-science-denial blog is run by a zoologist named Susan Crockford. It is called, without irony, *Polar Bear Science.*

Crockford grew up in Canada, where she developed an interest in the Arctic at a young age. She got her first malamute, the largest of the northern sled dogs, at age eleven and fostered an affinity for snowbound animals into her early adulthood, visiting the polar bears at the zoo in Vancouver. "As I watched the bears pace or play in their pool, my mind churned over the amazing phenomenon of seals and bears adapting to a life on sea ice—and wondered how a polar bear came to be in the first place," she wrote of her youth. Crockford attended the University of British

Columbia, the same school that legendary polar bear researcher Ian Stirling had gone to, and received her bachelor of science in zoology in 1976. In the late eighties, she co-founded a firm specializing in the identification and analysis of animal remains. In 2004, she completed her doctorate at the University of Victoria, and shortly thereafter she got a job there as an adjunct professor in the Department of Anthropology. Much of her early work focused on evolutionary biology, and she continued to delve into the history of the polar bear—how they split from prehistoric grizzlies and emerged as a distinct species. In 2012, she started what she would become most known for, the blog *Polar Bear Science*.

"I've had quite enough of the obfuscation of facts and model-based extrapolations into the future with regards to polar bears," she wrote in her inaugural post. "Spare us the emotional media hype, icon-peddling and fear-mongering about the future—we'd just like some information about the bears!"

In many of Crockford's posts lie grains of truth. She argues that the number of polar bears is growing, which is true for some populations, but twelve of the nineteen groups of bears are still considered "data deficient." No one knows precisely how many bears are in those populations, nor do they know if they are trending up or down. Though nearly every organization that studies them agrees that the best estimate of the total number of bears is between 22,000 and 31,000, Crockford made her own "best guess" that there were probably almost 40,000 bears and possibly as many as 58,000.

Sometimes the media plays into the hands of people like Crockford, as was the case with the viral images of one particular bear taken by wildlife photographers and conservationists Cristina Mittermeier and Paul Nicklen in 2017. Skin hung off the emaciated bear's frame, and it was obvious that the animal was close to death as it fruitlessly looked for food, chewing on an old snowmobile seat and peering into discarded, rusty barrels on the shore of Somerset Island, in the Canadian Arctic. "I took photographs, and Paul recorded video," Mittermeier later wrote. "As the bear approached the empty fuel drums looking for food, I could hear my colleagues sobbing." Nicklen posted the video on Instagram. "This is what starvation looks like," he wrote in the caption. "The muscles atrophy. No energy. It's a slow, painful death."

The images of the starving bear went viral, and news sites around the world picked up the story. *The Washington Post* called the heartbreaking photos "a rallying cry and stand-in for a largely unmitigated environmental disaster." *National Geographic* picked up the video, added some somber music, and took the narrative a step further. "This is what climate change looks like" flashed across the screen while the bear slowly ambled across the tundra.

It was impossible, of course, to know whether the bear's condition was caused by climate change. While the animal was very clearly on the verge of death from starvation, dying from a lack of nutrition is not uncommon among wild bears. When a bear gets injured, or too old to hunt, it will either be killed by another bear or eventually waste away. For Crockford and others inclined to doubt the effects of sea ice loss on polar bears, it was a golden opportunity. "One starving bear is not evidence of climate change, despite gruesome photos," read the headline on Crockford's blog. She accused the photographers of engaging in "tragedy porn" and decried the lack of any evidence that the bear starved because of climate change. She pointed out that if sea ice loss were the cause of this bear's unfortunate condition, it could be expected that many more bears in the area would be facing a similar fate, and none were documented. "One starving bear is not scientific evidence that man-made global warming has already negatively affected polar bears," she wrote, "but it is evidence that some activists will use any ploy to advance their agenda and increase donations."

Crockford was right in that one anecdotal photo, however gut-wrenching, does not in and of itself prove that polar bears across the Arctic are starving. For her part, Mittermeier freely admitted that the narrative had gotten away from her and Nicklen and morphed into something they couldn't control. While Nicklen alluded to the fact that climate change would eventually lead to the destruction of polar ecosystems, his Instagram caption only said, "This is what starvation looks like," without directly linking that individual bear's state to climate change. The intention, Mittermeier would later say, was to show what fate could await many more polar bears if sea ice continues to shrink. "Perhaps we made a mistake in not telling the full story—that we were looking for a picture that foretold the future and that we didn't know what had happened to this particular polar bear," she wrote for *National Geographic*. The magazine eventually changed the caption in the video and admitted that it had overplayed the role of climate change for this particular bear.

Crockford's blog doesn't exist in a vacuum. Through a process of citation and hyperlinking, denier blogs create their own echo chamber in which Crockford is promoted as "one of the world's foremost experts on polar bears," despite the fact that she has never conducted any field research and her views run contrary to nearly every mainstream polar bear expert. An analysis of climate denial blogs, conducted by some of the very scientists she frequently attacks, found that roughly 80 percent of them referenced Crockford's blog as their primary source when questioning the findings that polar bears are in trouble. The paper questioned Crockford's expertise and heavily implied that she uses tactics common among science deniers, like employing "rhetorical devices to evoke fear and other emotions, such as implying that the public is under threat from deceitful scientists." Crockford responded in a series of posts at *Polar Bear Science* calling the study "academic rape" and demanding a retraction from the journal that had published the study. The journal corrected two errors, one concerning the basis of Crockford's research and another on the sources of her funding, but stood by the overall analysis. In 2019, Crockford's contract at the University of Victoria was not renewed, a decision that she said was intended to silence her criticism of mainstream science. The university said it had nothing to do with her views. Crockford did not respond to a request for an interview.

Polar bears are charismatic, containing a number of seemingly contradictory qualities. They are endearing and ferocious. Strong as individuals but fragile as a species. They are to be feared, but also feared for. They come from a part of the world that few will ever see with their own eyes. It's no surprise they were anointed the face of climate change. Few people will be motivated to make changes to their lifestyle to save *Calanus glacialis,* the nutrient-rich Arctic copepod that ranks among the lowest links of the food chain.

But for all their cachet, polar bears make easy targets for deniers, too. Given the remoteness of their habitat, the status of many populations is uncertain. While some appear to have grown over time, that growth is more likely the product of hunting bans than of any change in ice conditions. But where there are knowledge gaps, deniers will seek to fill them with doubt. While zoos continue to promote their bears as captive representatives of their wild counterparts, a simple way to bring zoogoers face-to-face with climate change, the state of the animals doesn't easily adhere to a simple narrative. Even though it's widely agreed upon that their long-term survival is threatened by sea ice loss, their status is nuanced and complex. As an ambassador, the polar bear is imperfect.

A week after meeting for the first time, Nora and Hope were still struggling to get to know each other. Nora kept a wary eye on the larger bear and never got too close. All of Nora's moves, from Ohio to Oregon to Utah, had been in service of finding her a companion from whom she could learn. Now that she had one, the two were barely interacting. Hope had all the traits of an adolescent polar bear, but Nora still likely thought she was a person. Randinitis wasn't sure it was a hurdle they would ever overcome. As fall turned to winter in Salt Lake City, she began to think that maybe Nora and Hope weren't the best match, that maybe the two bears with very different upbringings would tolerate each other but never fully accept each other socially.

Then the keepers began to notice subtle changes in Nora's behavior.

She kept her distance from the larger bear, but sometimes while she was drinking water she would allow Hope to get close, pretending she was unaware of the approach. At first they took turns getting in the pool, but within a few weeks they had both become comfortable with swimming at the same time. Hope would stalk along the berm at the back of the exhibit, surveying the yard from its highest point, and Nora would follow along from below, watching her every move. When Hope would turn toward her smaller counterpart, Nora was off, scampering back to the safety of a corner.

Around Nora's second birthday in November, it became obvious to Randinitis and Jablonski that the young bear, for all her skittishness, was learning. Hope would walk up a log in a specific way, then, as soon as she was across the exhibit, Nora would repeat the action, step for step. When Hope twisted a ball to get at the food inside, Nora would be right behind her, doing the same. Hope would rub her body up against one of the logs, then casually walk away, while Nora peered over the berm, watching intently. Moments later, Nora was up against the same log, muzzle pressed to wood. It went both ways, too. The two bears traded off roles as pursuer and evader, and it seemed to Randinitis that Hope noticed how Nora interacted with the people on the other side of the glass. Soon the larger bear was up against the windows, inspecting the strange creatures looking back at her.

Nora maintained a bubble of personal space, but it was shrinking. In mid-December, almost ten weeks after she met Hope, the bubble popped.

Randinitis and another keeper were on the public side, watching the bears through the glass with the rest of the zoo-goers. The other keepers were in the back area, going about their daily tasks. Suddenly, the bears were right up next to each other. Hope offered a playful swipe at Nora, who responded with a lunge, aiming for but missing the bigger bear's neck. After a few seconds of open-mouth posturing, they were both on their hind legs, arms wrapped around each other, waving their heads back and forth, showing off their teeth. Hope could have easily overpowered her smaller counterpart, but instead she sat down and let Nora roll her over, the yellow-gray of both bears' fur blending in with the dirty dusting of snow in the exhibit. Nora hopped on top, and the two slapped at and bit around each other's faces.

Randinitis was ecstatic. To the untrained eye, it might have looked like Hope and Nora were fighting, but there was nothing aggressive about what the bears were doing. The submissive role that Hope had shown signs of was on full display. To Randinitis, it looked like they were having fun. As soon as she realized what was happening, she hopped on the radio and relayed the news. The bears had made the breakthrough everyone was waiting for. Nora and Hope were playing.

The response over the radio was a choir of excited zookeepers yelling *"Yay!"*

Chapter 18

Broken

After that initial contact, it was as if a switch had been flipped between Nora and Hope. Randinitis liked to imagine that the two had engaged in a late-night conversation in the holding area to work out whatever was keeping them apart.

Whatever had caused the change, the result was a blossoming companionship. Nora and Hope wrestled on land and slap-boxed in the pool, dunking each other underwater with their massive paws. They chased each other, Hope climbing onto an outcropping next to the water, Nora right behind her. When Hope jumped in, Nora would quickly follow. They would play-fight in the shallow end, standing on their hind legs, bodies towering out of the water. Sometimes they wrestled in the deeper end of the pool. From the other side of the glass, their writhing bodies almost appeared to become one mass of rippling white fur, rolling and turning in the water.

The deep end of the pool had two windows, one underwater, the other above, that looked into the neighboring sea lion enclosure. Diego and Maverick, the pinnipeds who lived there, were a constant source of fascination for Nora. With the exception of a few live fish in Columbus, Nora had never hunted anything in her life. Sea lions wouldn't have been her natural prey anyway. The farthest north most sea lions go is southern Alaska, well south of the southernmost wild polar bears. But given their re-semblance to seals, it's not hard to imagine that Nora's instincts—passed down from her wild grandmother—had something to do with her interest in her next-door neighbors. Diego and Maver-ick would swim endless laps, their sleek bodies gliding past the windows as Nora perched by the glass, transfixed. On one such occasion, Hope swam up behind Nora, who was resting on the windowsill. In one stealthy move, Hope approached from under-water and playfully chomped down on her smaller counterpart's hindquarters. Nora's head craned back in surprise and she lunged into the water after Hope. Those types of shenanigans became commonplace when the two bears were together.

Shannon Morarity, the Nora Mom from Columbus who teared up on TV on Nora's first day in front of the public, came to visit just after a massive storm had dumped nearly a foot of snow on Salt Lake City. She approached the glass tentatively, not sure that the bear she'd raised as a tiny, blind cub would recog-nize her after nearly two years. Morarity waited until the Hogle keepers started tossing fish into the enclosure. When Nora got up to wander toward her meal, Morarity ran up to the glass. Nora looked to the window, then down at the water in front of her. She looked up one more time at her former caretaker, and Morarity felt like something clicked. Their eyes locked and Nora came to the window, sniffing while Morarity held her hands up to the glass. It was as close as she could get to the creature she had bottle-fed as a newborn cub, when Nora still fit in her hands.

With Nora and Hope making good progress, the zoo staff focused more intently on Nora's health. Nearly all of the ste-reotypic behaviors that had troubled their counterparts in Portland—the pacing, the tantrums, the shunning of keepers—had ceased. Erika Crook, the zoo vet in Utah, weaned Nora off Xanax, and she responded positively. They started weaning her off Prozac as well, and that went smoothly. By the spring of 2018, Nora was free of mood-altering medications for the first time since the late fall of 2016.

Her bones, however, were still an area of concern. Nora continued to walk with a "bulldog stance." In February, Crook compared X-rays of Nora's elbow taken in Portland over the previous summer with images of Hope's bones. The difference was stark. On Nora's left arm, a large, round mass of bone protruded from the humerus, the long bone that connects the shoulder to the elbow. X-rays of the same bone in Hope showed no such lump. The vets at Hogle used an infrared camera to take thermal images of Nora. The camera, which shows heat signatures from inflammation, was the least invasive way they could get an idea of what was going on with her joints. Her right side showed cool colors, greens and blues that denoted health in the tissues beneath her skin. Her left side was ablaze with yellow, signs that there was swelling in the area, and blotches of red marked her front elbow and shoulder, suggesting early-onset arthritis. Crook kept up her regimen of anti-inflammatory medications and painkillers, a handful of pills administered daily, often in a mix of Cheerios and peanut butter.

Nora showed few signs of pain as she continued to build a relationship with Hope, romping in the pool and play-fighting with her companion for the toys donated to the zoo by Salt Lake City residents ahead of their arrival. Crook hoped the young bear would maintain the trajectory she was on, tolerating her misshapen bones. But polar bears can't tell you when they are in pain, and the full extent of Nora's bone problems wouldn't be known until she finished growing. There was a chance, Crook knew, that Nora's bones could degenerate further, and her pain could rise to a level where the zoo would be forced to make difficult decisions about her quality of life.

As summer faded into fall, those chances grew more remote. In the Hogle enclosure, built to resemble the natural world, Nora appeared to be thriving.

Winter got off to a slow start in the Bering Sea in 2017. Elevated air and sea surface temperatures lingered from that fall into the new year, and strong south winds kicked up more storms than usual, drawing warm air into the Arctic. The ice that did form was thin and easily broken up. Floating ice moved in and out of the region, and the back-and-forth of icebergs passing through the Bering Strait made boat travel dangerous. Shorefast ice, a necessity for hunting and fishing and traveling by snowmobile, was slow to form. In Shishmaref, there was open water around the island until January. The community has historically been iced in by the end of October.

In late February, residents of Little Diomede noticed that the rising tide was fragmenting what little ice clung to the shore of the island. A few days later, strong winds heralded what would be the strongest of ten storms to pass through the region in a two-week period. There's no weather station on Little Diomede, but the one twenty-five miles to the east in Wales clocked gusts of eighty-six miles per hour on February 20. A National Weather Service meteorologist working in Fairbanks told the *Anchorage Daily News* that he issued a flood warning for Diomede before the storms hit, something he'd never done in February in his thirty years on the job. Huge waves crashed well above the high-tide line, and the ocean hurled large chunks of ice and debris onto land. A hunting boat, one of three in the village, was destroyed. Ice and seawater got into the water treatment building. Electrical lines were knocked out by wind and salt spray. The crashing waves also damaged the helipad, which at that time was the only way on or off the island. Thankfully for Diomede's residents, the helipad was still functional, but the incident cast the vulnerability of the island's critical infrastructure in stark relief. Usually at that time of year residents have already set up a runway for small planes on the sea ice; in 2018, the runway site was open water.

The story was similar farther south on the St. Lawrence Island village of Gambell. Ice-free waters stretched all the way to Russia into March. After the economic disasters of 2013 and 2015, the walrus hunters of Gambell again reported a shortage of game. In April, just over 38,000 square miles of ocean were covered with ice—as compared with 400,000 in 2013, a year that hewed closer to the thirty-year average. There hadn't been so little ice in the Bering Sea since white sailors on whaling ships began keeping official records in the 1850s.

The problems weren't confined to Alaska. Though some parts of the Arctic were near average or a little below, many regions saw record-breaking high temperatures. In February, the closest weather station to the North Pole, at Cape Morris Jesup, on Greenland's north coast, observed sixty-one hours above freezing for the month, including one period of nearly twenty-four hours. At one point, the temperature at the top of the world was forty-five degrees above normal for that time of year. Some climatologists believe this happened because the jet stream had weakened, allowing warm air to creep north and displacing the polar vortex, a low-pressure system that usually keeps the Arctic frigid, to lower latitudes. While the North Pole was forty-five degrees above normal, much of Western Europe was enduring a blast of Arctic weather. Freezing temperatures stretched from Spain to Poland. Rome saw snow for the first time in six years.

Record high temperatures lasted into the summer, both in the far north and elsewhere. In July, several Scandinavian countries, where summer is normally temperate, saw highs above ninety degrees. In Sweden, the country's tallest peak became its second-tallest peak after nearly two feet melted off the glacier that capped its summit. The heat was not confined to the north, though. Los Angeles hit 111 degrees in July, the same month that Ouargla, Algeria, saw the hottest temperature ever reliably measured on the African continent, at just over 124 degrees.

Wildfires, in most cases started by humans but fueled by hot temperatures and a dearth of rain, blazed through Greece as fifty-mile-per-hour winds fanned the flames. Thousands were evacuated, and some, with escape routes blocked, were forced to flee the blaze by jumping into the Aegean Sea. More than a hundred people were killed. Britain and Sweden both experienced widespread forest fires, too.

In November, well into what should have been the rainy season in Northern California, a perfect storm of fire conditions was brewing in rural Butte County, about 150 miles north of San Francisco. After years of drought and an unusually dry summer, winds began to build out of the northeast, gusting up to fifty miles per hour through the Sierra Nevada foothills. Around 6:30 A.M., an electrical transmission line cast a spark into the dry brush near Camp Creek Road, in the Feather River Canyon. Within twelve hours, the Camp Fire had burned 55,000 acres. Nearly 19,000 structures burned, most of them houses, leaving around 30,000 people homeless. Eighty-six people lost their lives, the vast majority of them residents of the town of Paradise over the age of sixty. It was the deadliest wildfire in California history.

Heat records are set every year. Droughts come and go. Wildfires burn and winds blow, and the two combine to scorch hundreds of thousands of acres. The natural world has always worked that way. But the way the Camp Fire burned, while the rest of the world simmered under record-breaking temperatures, had all the fingerprints of climate change.

Earlier that year, a state agency in California released a report detailing how it planned to fight wildfires like the ones that had devastated the state two years in a row. "Climate change has rendered the term 'fire season' obsolete, as wildfires now burn on a year-round basis across the State," it read.

Donald Trump announced his intent to pull the United States out of the Paris Agreement five months into his presidency, and though it was his most publicized rebuke of climate science, it was far from the only one. Throughout the first part of the Trump administration, the federal government stayed busy rolling back all kinds of environmental regulations, large and small.

Under the previous administration, the Environmental Protection Agency had sent out more than fifteen thousand surveys to oil and gas production facilities, asking the operators to provide information on their emissions. Trump's newly installed EPA administrator, Scott Pruitt, who had come under fire during his confirmation hearings for his close ties to fossil fuel production, canceled the policy, saying he was aiming to "reduce burdens on businesses." The Obama administration had instituted a wonky-sounding cost-benefit analysis tool called the "social cost of carbon," meant to measure the benefits for future generations from even modest greenhouse gas reductions. Trump rescinded it with an executive order, in order to "promote clean and safe development of our Nation's vast energy resources, while at the same time avoiding regulatory burdens that unnecessarily encumber energy production." There had been a freeze on new leases for coal mining on public land; Trump lifted it. In 2016, Obama had created what was called the

Northern Bering Sea Climate Resilience Area, the product of work with locals, nonprofits, and Native Alaskans to protect the waters around Nome and Wales and Shishmaref from fossil fuel exploration. Obama's executive order had also created an advisory council, one that included tribal representation, to offer input on federal decisions that would impact the region. Trump rescinded that, too. He withdrew an order telling national parks to consider climate change when managing their resources. He repealed, rolled back, rescinded, or signaled his desire to reverse nearly eighty environmental regulations by the end of 2018.

In October of that year, the United Nations Intergovernmental Panel on Climate Change released its most dire report yet, warning that the planet would likely see catastrophic effects of climate change even if it managed to stay under the two-degree Celsius warming threshold set by the Paris Agreement. What we must strive for, the ninety-one authors from forty countries wrote, was a lower benchmark of 1.5 degrees Celsius. That half degree would be crucial, they said, in averting widespread extinction of insects and animals, tempering the impacts of a warming world on food production, and keeping the permafrost near the world's poles from melting. With 1.5 degrees of warming, 14 percent of the world's population would be exposed to scorching temperatures like the ones seen around the world in 2018 every five years. With two degrees, that number would jump to 37 percent. The lower threshold of 1.5 degrees would translate to an effectively ice-free Arctic summer one out of every hundred years. Two degrees would make that one out of every ten.

Six weeks later, the U.S. government's Global Change Research Program released its own report, the fourth in a series of congressionally mandated publications meant to perform a similar function to the IPCC's reports but on a national level. Climate change was already here, its effects were already being felt, and the need to take action was urgent, the report said. But even as the impacts of global warming were becoming more apparent, it was becoming equally clear that the brunt of the burden would fall on the shoulders of those least able to bear it. These "front-line communities," as they were described in the National Climate Assessment, are the poor, the vulnerable, and those that depend on natural resources for their survival. They are the elderly, who perished by the thousands during Europe's 2003 heat wave. They are the Native people of Shishmaref, watching as storms eat away at the only home they have ever known. They are homeless people in Portland, who have nowhere to go when the city is blanketed in toxic smoke and wildfire ash.

Climate change would come for them first, the report said, but it would eventually come for us all.

As winter settled into Utah, Nora and Hope were showing signs of true companionship. Both bears were still growing: Nora was around five hundred pounds by the fall, and Hope had nearly a hundred pounds on her. Still, the larger bear had taken on an almost maternal role, teaching Nora how to play by rolling on her back when they wrestled. They slept near the window sometimes, not in actual physical contact with each other, but in as close proximity as most polar bears will allow. As 2018 rolled into 2019, Randinitis and Jablonski and the other keepers couldn't have been more pleased with how the bears' relationship was developing.

Then, on a Thursday in late January, Randinitis was at home on one of her days off when she got a text from a colleague. Something was wrong with Nora. Staff who had done the morning walk-through had noticed that Nora was in one of the shelters in the exhibit. She hadn't come when she was called, nor when Hope came out into the yard. She had her paws over her eyes and looked tired, one of the keepers wrote in their daily logbook. By midmorning, she was still in the same position. She picked up her head briefly when they tossed her some fish, but she lay back down almost immediately and stayed like that for the rest of the day. Zoo staff who monitor the grounds at night were told to keep an eye on her, but nobody saw her move from the den.

The next morning, Friday, Nora was still there. Erika Crook, the zoo vet, began to worry that Nora had eaten something she shouldn't have. She hadn't shown any symptoms of stomach problems, no missed meals or vomiting. But an intestinal blockage can be fatal for polar bears, so it became clear that they would need to examine her. There was no way to do so while she was in the public-facing enclosure, though. In theory, they could have darted her with a gun, but if she wandered too close to the pool before the sedatives took effect, she could fall in and drown. The keepers and vets had to wait, but their concerns were growing. They tried to call Nora over for training, for which she was usually an enthusiastic participant, but she didn't budge. They called her from the catwalks that overlooked the enclosure and from the door to the holding area, but Nora, who almost never missed an opportunity to interact with her keepers, was steadfast. At one point on Friday, she stood up when keepers tossed in some more fish, but her back end looked wobbly and she was favoring her right arm. She stood for only half an hour, though, and then went back to lying in the den.

When Randinitis came in on Saturday, Nora was in the demo room, a covered area halfway between the enclosure and the holding area. There's a bed in the room, a large piece of heavy-duty canvas hung between four posts, like an industrial hammock, and Nora appeared to be stuck in it. The keepers looked at video footage from the night before and saw that she had gotten into the room by scooting backwards across the yard, never putting any weight on her front limbs, essentially dragging the top half of her body. It was behavior none of them had seen before from Nora—or any bear. Over the course of an hour, they coaxed her back to the holding area and got her into one of the stalls, where she lay down, seemingly exhausted. Keepers coated some painkillers in peanut butter and tossed them to her. She ate a few and ignored the rest while the vets got a better look at her. She wasn't bending her right arm, and it looked to be swollen to nearly twice its normal size. Given her history, though, they weren't quite sure what the problem was. It could have been her arthritis flaring up or a sprained wrist or a dislocated joint. They needed to take X-rays.

On Sunday, Crook—along with another vet and a technician who were supposed to have the day off but came in to help— sedated Nora and brought in the zoo's portable radiograph machine, a yellow contraption originally intended to diagnose horses. It was connected by a long cable to a laptop nearby. Jablonski and Randinitis were both there, equal parts curious and worried.

When the first images were transmitted, it immediately became clear what was wrong. Nora's right humerus—the one opposite the bone that had shown a lump the previous spring— was broken. One of the biggest weight-bearing bones in her body looked like it had been snapped in half, each end pointed like a dry stick that had been bent too far.

Jablonski felt an odd sense of relief. It was a horrible injury, but now at least they knew what they were dealing with. When Randinitis saw the image of the bisected bone, however, it felt to her like all the air had gotten sucked out of the room. Her mind immediately went into triage mode, the questions coming fast and furious. How would they fix this? Would they have to amputate her leg? What kind of life could a three-legged bear even live? She'd seen the skeletons of wild animals that had suffered broken bones and knew that animals sometimes have astonishing abilities to heal themselves. But in the wild, animals also suffer and die. That's not how zoos operate.

Chapter 19

A Risky Repair

Jeff Watkins had spent more than thirty years, essentially his whole career, fixing fractures in large animals, but the email he got in late January still caught him by surprise.

Watkins had studied veterinary medicine in Kansas before going to grad school at Texas A&M, where he became fascinated with the anatomy of horses—the interplay of muscle and bone, how joints rotate and support the mass of animals that weigh hundreds of pounds. More specifically, he was interested in how to fix fractures in large bones. Minor breaks could sometimes be allowed to heal in a cast, but for severe breaks, a cast is not an option—the animal has to undergo surgery. And performing surgery on a horse isn't easy. Foals need to be able to put weight on all four legs almost immediately after the procedure, or else complications can occur—their other legs might develop joint issues, muscles can atrophy, tendons may contract. In the mid-eighties, when Watkins was training in orthopedic surgery, most foals that broke their legs had to be euthanized.

Doctors have experimented with using metal implants to fix broken bones in humans since at least the early 1890s, and by the 1970s they had begun using a type of rod with holes into which screws could be inserted perpendicular to the bone, above and below the site of the fracture. The addition of interlocking

screws prevented the segments of the bone from rotating or tele-
scoping over each other and allowed the injured limb to bear
weight soon after surgery. By the early eighties, the interlocking
intramedullary nail was a common treatment for people with
broken femurs and tibias.

After creating his own custom tools to complete the procedure, Watkins pioneered the use of the intramedullary nail in
young horses. He traveled all over the country performing the
surgeries, including to the veterinary school at the University of
California, Davis. He met another surgeon there, Amy Kapatkin,
who had operated on the broken femur of an eighteen-month-
old polar bear named Tundra at the Bronx Zoo in 1993.

When the vets at Hogle saw the images of Nora's humerus,
they knew that the type of surgery required to fix a bone of that
size was beyond their expertise. They started locally, calling a
small animal surgeon in Salt Lake City. He was interested but
didn't have the right equipment. The zoo vets expanded their
search, calling and sending emails to anyone they thought might
be able to help. One person had heard of Tundra and referred
Erika Crook to Kapatkin at UC Davis, who in turn referred her
to Watkins at Texas A&M. That's how the horse surgeon ended
up getting an email in late January asking if he'd be willing to
operate on a very broken polar bear named Nora.

Watkins had never operated on a polar bear, and nobody he
knew of had ever attempted to fix a humerus in one. Nora's leg
had already been broken for several days at that point, and the
longer it remained untreated, the harder it would be to repair.
With every day that passed, Nora's muscles stiffened more
around the broken bone. If he was going to attempt the surgery,
there would be a very small window in which to arrange all the
logistics and complete the procedure.

He'd consider it, he told Crook, but only if he could bring

along another veterinary orthopedic surgeon from Texas A&M named Kati Glass. She knew the procedure and was familiar with the equipment, and, most important, Watkins trusted her. She began researching polar bear anatomy while Watkins set about tracking down the equipment he would need. He had the interlocking nail, but for an animal of Nora's size, he wanted to use a plate, too, for added strength, and he'd need to have it sent to Utah. He would also need a specific type of drill, as well as specialized screws and various sizes of medical drill bits. And he would need it all to arrive in Utah, sterilized, in a matter of days. He called the rep at his medical supplier, Johnson & Johnson, and they donated $90,000 worth of equipment, promising it would be where it needed to be by the time Watkins was ready to operate. He got the X-rays of Nora's fracture from the vets at Hogle, but he couldn't tell how big the cavity in the bone was and he wasn't sure the rod would fit.

Watkins and Glass had never heard of Nora, but in the course of her research Glass learned that their prospective patient was a celebrity in the zoo world. Her dedicated fans in Oregon and Ohio who would be satisfied with nothing less than a completely successful operation. Her fame added another layer of pressure to the surgery. When he operates on horses, Watkins usually has a frank discussion with the owner beforehand. He tells them that sometimes, during the course of the operation, it becomes clear that the fracture is beyond repair. If he can't get the bone fragments to align just right or if the repair will leave the foal in pain for the rest of its life, the only humane option is to euthanize the animal. Watkins knew he couldn't consider that as an option for Nora.

An anesthesiologist was brought in from North Carolina, and a local orthopedic surgeon was called in to assist. The equipment that had been shipped to Utah arrived on time and undamaged. The following Sunday, Watkins flew to Salt Lake City. He would operate the next day, more than a week after Nora had been discovered not moving in her enclosure.

Early Monday morning, a team of vets and keepers sedated Nora in the holding area. She was carried the short distance to the zoo hospital on a cargo net hoisted by ten people, including Randinitis and Jablonski. For Crook, the whole first hour was a whirlwind of activity, but the move was one of the most stressful times. She had been responsible for sedating Nora, and she kept a close eye on the bear for any signs that she was waking up. She was injured and unconscious, but Nora was still a five-hundred-pound carnivore, and now she was essentially uncontained. If she woke up, the situation could get dangerous very quickly. But the sedatives worked, and the sleeping bear made it to the operating table without incident.

Nora was positioned on the table while Crook assisted the anesthesiologist. Watkins and Glass and the other members of the surgical team scrubbed in as Nora was intubated and put on a ventilator. Crook helped attach a catheter to the bear's ear so they could monitor her blood pressure and other vital signs while Jablonski and Randinitis stood on the other side of a window. Watkins was doing his best to focus on the task at hand, but there was a lot to marvel at. As they shaved the hair from the incision site, he was shocked to realize that under all that white fur, polar bears have dark, almost purple skin.

Nearly an hour after she was sedated, Nora was fully prepped and covered in surgical drapes, and the procedure started in earnest. Watkins cut a nearly fifteen-inch incision, which curved around Nora's arm from her shoulder to just above her elbow. The team peeled back her skin and began making their way toward the site of the fracture. It wasn't so different from a horse, Watkins thought to himself. Once they got to the break itself,

Watkins knew it was going to be difficult to get the bones lined up, which is necessary for the rod to be inserted correctly. He began trying all the things he'd learned on horses over his career, but nothing was working. It took a lot of effort, both mentally and physically, and sweat beaded on Watkins's brow. Out of other options, he tried to tent the bone, bending the two segments up into a triangle in hopes that he could pull them back down into a straight line, but no amount of force could overcome Nora's muscles.

In November of 2016, around the time Nora was celebrating her first birthday, Jamie Margolin was in her hometown of Seattle, contemplating the aftermath of Donald Trump's election. She was an only child and wasn't one of the popular kids in school, so she'd always been self-motivated, finding ways to entertain herself with art projects and writing. She'd been concerned about climate change and involved in politics for a few years, volunteering at the local Democratic campaign headquarters ahead of the 2016 election. Margolin had always figured that if the right people were elected, leaders who were intent on taking meaningful action on climate issues, change would follow. When Trump won the presidency, she realized that change was still a long way off.

The next summer, as Hurricanes Harvey and Maria ravaged Texas and Puerto Rico, smoke from wildfires in Canada enveloped Seattle. Margolin, sixteen at the time, decided she needed to take matters into her own hands. She had seen the power of the Women's March, when millions of people took to the streets the day after Trump's inauguration, and she grew fascinated with mass mobilization. She wrote a post on social media asking if anyone she knew would join her in a climate march, and soon her direct message inbox was lighting up with people who wanted in. She co-founded Zero Hour, one of dozens of youth groups that have emerged as the dire consequences of climate change have come into sharper relief.

Margolin identifies as a Colombian American Jewish lesbian, and social justice is at the core of her group's environmental advocacy. Zero Hour identifies systems of oppression—racism, sexism, patriarchy, and capitalism, along with the same type of colonialism that led missionaries to Wales in the 1890s—as among the root causes of the climate crisis. To build a society that doesn't rely on fossil fuels, the group advocates for a "just transition," one that takes into account the needs of marginalized communities, the same ones that are most likely to feel the worst effects of a warming planet, while also looking to those frontline communities for solutions. In 2018, like Esau Sinnok in Alaska and Kelsey Juliana in Oregon, she joined a lawsuit against the State of Washington.

That summer, while Nora was playing with Hope in Utah, Margolin was helping to plan a series of events in Washington, D.C. The first day, more than a hundred young people went to the Capitol, delivering a list of demands to lawmakers. They called on Congress to declare a climate emergency, eliminate federal subsidies to fossil fuel producers, and rejoin the Paris Agreement. They also demanded that the country recognize Indigenous treaty rights and ensure that legislation addressing the climate emergency not put disproportionate burdens on communities of color and low-income communities. They called for a ban on new fossil fuel infrastructure by 2030 and a massive investment in mass transit. And they called for a complete transition away from fossil fuels within twenty years, with a shift toward local economies that focus on sustainability. By then, they wrote, greenhouse gas emissions should be negative—any emissions offset by reforestation or technology that can suck carbon dioxide out of the atmosphere—with an eventual goal of reaching atmospheric carbon dioxide concentrations of no more than 350 parts per million.

"2040 is just 22 years away," the declaration read. "This is ZERO HOUR."

Two days after delivering those demands to lawmakers, on a rainy Saturday in July, hundreds of young people marched through the rain, led by Indigenous leaders from the protest at the Standing Rock Indian Reservation, in North Dakota, where hundreds of Native Americans had stood against the completion of an oil pipeline in 2016.

A few weeks after the Zero Hour march, Greta Thunberg, a Swedish teenager, decided to skip school to protest her government's inaction on global warming as the national elections approached. She'd been concerned with climate change for years, giving up meat and air travel at the age of twelve to lessen her carbon footprint, but the heat wave and wildfires that had swept across Sweden that summer had given her a sense of urgency. On August 20, she posted on Instagram a picture of herself sitting on the ground outside Sweden's parliament, in Stockholm, with a sign that read, SKOLSTREJK FÖR KLIMATET, or "School Strike for Climate."

"We children don't usually do what you grown-ups tell us to do. We do as you do. And since you don't give a shit about my future, I don't give a shit, either," she wrote in the caption. "I strike for the climate until Election Day."

Within a week, thirty-five people had joined the strike. After the election, in September, Thunberg went back to school for four days a week, but she returned to protesting the next Friday and the Friday after that. Her posts spread across social media under the hashtag #FridaysForFuture. By October 20, days after the IPCC released its alarming report on the 1.5-degree thresh-old, Thunberg was speaking to a gathering of ten thousand marchers in Finland. In December, the fifteen-year-old addressed the United Nations Climate Change Conference in Poland. In January 2019, days before Nora broke her leg, Thunberg spoke at the World Economic Forum, in Davos, Switzerland, asking the attendees why there was no sense of panic among them. "Our house is on fire," she told the crowd. "You say nothing in life is black or white. But that is a lie, a very dangerous lie. Either we prevent 1.5 degrees of warming or we don't. Either we avoid set-ting off that irreversible chain reaction beyond human control or we don't." Thunberg had been diagnosed with Asperger's a few years before her first climate strike, and she sees it as a super-power rather than a disability. Her ability to speak to the world's most powerful men in a direct and candid manner won her fans and followers around the world and made her the figurehead of the youth climate movement.

After the Democrats won a majority in the U.S. House of Rep-resentatives in the November midterm elections, more than 150 young people from a group called the Sunrise Movement staged a sit-in outside House Speaker Nancy Pelosi's office, demanding she support a loosely defined but wide-ranging suite of climate regulations called the Green New Deal. They were joined by newly elected progressive congresswoman Alexandria Ocasio-Cortez, a young Latina woman who had just unseated the in-cumbent in her home district in New York. Ocasio-Cortez began lobbying Pelosi to take action on climate change by pressing for the Green New Deal. The framework was easily among the most ambitious climate proposals put forward by any U.S. lawmaker, one that not only called for massive changes to energy produc-tion and investment in green technology but contained within it a strong social justice component.

The plan was an opportunity to reset, and the young people who advocated for it—Jamie Margolin and Greta Thunberg among them—had their eyes set on a future framed by equity. Equity across race. Equity across gender. Equity across all spectrums of society.

Just outside the operating room, Randinitis watched intently from behind the glass. There was nothing she could do from her side of the window, but she felt she needed to be there just the same. She never thought of the zoo's animals as pets, not fully, but in her mind they were like some combination of a pet, a child, and a co-worker. She felt it was necessary for Nora to know she was there, even if only on a subconscious level. Jablonski was there, too, watching as Watkins struggled to get the segments of Nora's humerus to line up. Nora wasn't the only animal in their care, so the keepers ducked out when they had to tend to their other charges. But keepers from other parts of the zoo volunteered to cover for them so they could watch as much of the procedure as possible.

When Watkins had been talking with the reps at Johnson & Johnson, they'd asked him what size bone saw he wanted. The surgeon hadn't included a saw in his initial order—he didn't normally use one for this type of procedure—but he knew that if he turned it down, he'd find himself needing one. They threw in a blade with the other equipment, and as Watkins weighed the risks of cutting into Nora's already broken bone, he was glad they had.

It was a measure of last resort. The goal of the surgery was to repair the bone and leave it intact. Removing any material could create complications; he'd never had to do it before, and he wasn't sure exactly how it would affect the healing process. But he'd already been working on Nora for a few hours at that point, and the longer an animal stays under anesthesia, the riskier the procedure becomes. He attached the saw blade to the drill and trimmed off a small pointy piece of the upper bone segment, the part that attached to the shoulder. It did the trick. The bones were brought into alignment, and Watkins put a big clamp on the humerus to hold it in place. He slid the nail down into the cavity and used a special tool to target the holes, inserting three screws above the site of the fracture and two below.

In a foal of Nora's size, he would have added a plate, too. But Nora had been in surgery for several hours and under anesthesia even longer. Her vital signs were beginning to falter, and the risk of infection increased with every minute she was open on the table. Putting in a plate would take another couple of hours at least, and Watkins didn't want to risk it. He wrapped the bone in cables that he cinched down and crimped in place, right around the break, hoping they would provide the fixation that Nora would need to heal. With a horse, Watkins would be able to dress the wound and manage it throughout the healing process. Once Nora woke up, no one would be able to touch her again without putting her under, and there was no way to bandage her wound or keep her from licking the incision site.

All they could do now was wait for Nora to wake up and hope for the best.

Chapter 20

Nora's Keepers

The next day, Nora was awake and moving around, albeit in a very limited way. The keepers kept her restricted to one of the small stalls in the rear holding area, with Hope sequestered on the other side of the exhibit. The bears had been separated since Nora's injury, but their powerful noses could smell each other. At first Hope hadn't seemed to mind being alone, but after a few days she realized that certain parts of the enclosure that she'd had access to were off-limits. She started banging on the doors, letting the keepers know she didn't appreciate it. It was impossible to know for sure if Hope was mad at being locked out of the holding area or if she was lonely, but it seemed to Randinitis as though she missed Nora.

Watkins came back to check on his patient the day after her surgery. She was obviously in some discomfort—she had just endured a long surgery in which her bones had been sawed apart and put back together—but her leg appeared to be structurally stable. He felt satisfied that they'd done their best to give her a solid chance at a successful recovery. It was hard for the keepers, though. Nora's rehab required keeping her in a confined space, gradually giving her more access as she healed. But Jablonski couldn't explain that to the bear, and as Nora regained some of her mobility, it was frustrating for the keeper to watch her limp around the enclosure, only gingerly using her right leg, clueless as to why she was being kept in a small room without access to the pool or the yard or her companion, Hope. The vets had prescribed trazodone, a sedative, to keep her calm, and painkillers to help with the pain at the incision site, all mixed into canned cat food that Nora eagerly lapped from a spoon the keepers stuck through the wire mesh of her pen.

Within a few weeks, white peach fuzz grew in where Nora's arm had been shaved, though it was still a deep purple from her shoulder to her wrist. Because the vets hadn't shaved her paw, it almost looked as if Nora was wearing a furry mitten. They sent pictures to Watkins, in Texas, and he was pleased with how the wound was healing. After about six weeks, the keepers gave Nora access to more space. She shuffled around the pen, putting more weight on her injured leg with each passing day.

Now that her surgery was over, there wasn't much the keepers could do to help her physically heal, but they did their best to keep her mind active. The zoo had been flooded with get-well cards, and Randinitis spent hours reading them to Nora. Keepers from all over the zoo took turns coming down to Rocky Shores to keep Nora company. The head of the commissary came down to read the newspaper to her. Jablonski read her emails and played music for her. She drew pictures with scented markers and let the bear investigate the strange smells. As the weeks wore on, the keepers worked with Watkins and Crook to develop a physical therapy routine, using fish to coax Nora into positions where she'd have to use her injured leg.

At ten weeks, they sedated Nora again for another set of X-rays. Images of the fracture showed a gap where Watkins had been forced to trim the upper segment of the humerus, but there were also signs of new growth where the bone was trying to bridge that gap. Everyone agreed that Nora was ready for a little more room to roam, so they gave her access to the smaller of the two outside yards, but they put a cover on the pool to keep her from injuring herself getting in and out. Finally, toward the end of May, nearly four months since her surgery, the vets gave Nora permission to swim. As soon as she saw that the pool had been uncovered, she made a beeline for it and promptly did a belly flop.

Still, it was clear that Nora missed Hope—and vice versa. The larger bear spent much of her time in the public-facing yard, but if she sat in the right spot, she could see Nora back in the holding area. When the two caught sight of each other, they would sit and stare for what seemed to Randinitis like forever. Sometimes, when Randinitis arrived in the morning, she'd find Nora and Hope sleeping on either side of the same door.

The shorefast ice was gone and open water lapped at the shores of Wales by the end of March, more than a month earlier than usual, leaving hunters like Gene Agnaboogok and Gilbert Oxereok a shorter window to search for the seals and walrus that followed the ice as it retreated north. It was troublingly similar to the previous winter, when the Bering Sea saw record low amounts of sea ice. As the temperature hovered just above freezing, heavy wet snow turned to slush. Frozen rivers in the southeast part of the state broke up early, and four people died when their vehicles crashed through river ice that was usually frozen solid at that time of year.

Climate deniers, skeptics, and contrarians continued their attacks on the science of global warming as the federal government cut regulations on a parallel track. The Trump administration mandated that future reports on climate from the U.S. Geological Survey, the agency that Karyn Rode works for, not provide any models beyond 2040, even though most scientists agreed that the worst impacts of a changing climate would come in the latter half of the century. Trump made official his pledge to withdraw from the Paris Agreement, filing the necessary paperwork to pull the United States from the pact the day after the 2020 election. At a meeting of the Arctic Council, an entity intended to foster cooperation among the world's eight Arctic nations, U.S. secretary of state Mike Pompeo noted that the diminishing sea ice was opening up previously untapped resources, including fossil fuels, to American exploration.

But there were signs that public sentiment was shifting. In March 2019, the Democratic presidential primary saw its first candidate to make climate change his signature issue, in Washington governor Jay Inslee. CNN hosted a seven-hour climate crisis town hall with ten presidential candidates. Poll after poll confirmed that climate change was among the top concerns of Democratic voters, even if Republican opinion was left largely unchanged amid increasing political polarization.

Young people continued to play a leading role in advocating for action. Greta Thunberg traveled to New York on a zero-emissions sailboat for a series of events centered on the United Nations Climate Summit. She went to Congress and testified alongside Zero Hour co-founder Jamie Margolin. "The fact that you are staring at a panel of young people testifying before you today pleading for a livable earth should not fill you with pride; it should fill you with shame," Margolin told the assembled lawmakers. "We are exhausted because we have tried everything." Instead of offering any prepared remarks, Thunberg submitted the IPCC's 1.5-degree report and told the lawmakers to "listen to the scientists."

Both were the subject of online harassment. Climate deniers made fun of Thunberg's braids and mocked her Asperger's diagnosis. They sent anti-Semitic slurs and veiled threats to Margolin, asking the Jewish teenager which synagogue she attended.

In between her speeches in Washington, D.C., and New York, Thunberg continued her Friday school strikes. When she had first skipped school to protest climate change, in August of 2018, she was alone. When she marched in New York just over a year later, she was one of millions of people around the world who had skipped school and work to demand action. There were more than eight hundred rallies in the United States alone. Thousands of students walked out of Portland classrooms and choked the streets down the hill from the Oregon Zoo. A small group gathered in Brevig Mission, the village farther down the cape from Wales. Demonstrators marched in Australia and India and Turkey. Hundreds marched in Columbus. In Salt Lake City, they marched in front of the state capitol, just miles from a young zoo bear who was still recovering from surgery. A bear whose father was born in the wild. A bear whose grandmother stalked the very ice the young protesters were demanding adults act to save.

The keepers never figured out how Nora had hurt herself. She hadn't taken any obvious tumbles or engaged in particularly vigorous roughhousing with Hope that anyone had seen. She had always played rough and had a habit of slamming up against the glass paws-first. The keepers' best guess was that Nora broke her humerus goofing off, perhaps landing on her leg at a bad angle after an especially forceful pounce.

Still, the broken leg raised fresh concerns about Nora's skeletal integrity. Had the metabolic bone disease she'd experienced as a cub left her bones brittle or soft? Crook didn't think so, but to be sure, they sent the piece of bone that Watkins had sawed off during surgery to a lab for analysis. It passed every test it was put through with flying colors. It had all the right vitamins and minerals. It was just as dense as polar bear bone should be. However the fracture occurred, it wasn't because of anything that had happened to Nora as a newborn.

By summer Nora was letting her keepers know she was ready for more space. She would sit by the door, looking outside, then looking at Randinitis or Jablonski, then repeating the process over and over, like a dog waiting to be let out. The groundskeepers at the zoo went into the enclosure to "babyproof" the space, looking for any areas that could be problematic for a still-recovering bear. They filled in holes and took out a tree stump. They used gravel to close gaps between large rocks where a bear could potentially catch an ankle. They trimmed one of the bigger cottonwood logs. By August all of Nora's hair had grown back, the long, curving surgical scar visible only after she swam and her fur became matted to her body. Her rehab had progressed as well. She was putting her full weight on the repaired leg and seemed to have no problems moving around. Her limp was nearly imperceptible.

Six months removed from her procedure, the vets at Hogle took another set of X-rays and sent them off to Watkins. There was good bone growth and the rod was working as intended, though he would have expected a foal to be a little further along in the healing process than Nora was. Then again, he had no reference point for how long a polar bear humerus takes to heal. By the end of the summer, Nora was back in the public-facing exhibit, meandering up the logs, swimming in the pool, and floating in her favorite spot by the windows, the one that looked into the pinniped pool next door, watching as the sea lions glided past. As the weather began to cool in Salt Lake City, Watkins and Glass traveled back to see Nora. They did another round of X-rays, and it looked to the surgeons like Nora would make a full recovery.

In November, Nora turned four years old. The zoo marked the occasion with a luau-themed party. A crowd of children sang "Happy Birthday" as Nora chomped into a cake made of fish fat, oatmeal, apricot juice, and salmon. She and Hope were still being kept separate, the keepers waiting until they were as sure as they could be that rough play wouldn't reinjure Nora, who now weighed 550 pounds, to Hope's 700.

Nora's joints would never be perfect, and the spots where her shoulders and elbows connected would never be as clean and crisp as in other bears. She would probably always walk with her trademark bulldog stance. But it didn't seem to impinge on her quality of life. Nora was doing well enough that the vet decided to try weaning her off some of the anti-inflammatory meds she'd been taking since her time in Portland. She still suckled on the mesh of her enclosures on occasion, but all of her other stereotypic behaviors had ceased. Because of the monumental work put in by Randinitis and Jablonski and Crook and Watkins, along with a host of other vets and keepers, Nora's prognosis was good, which was more than could be said for her wild counterparts.

Hunters killing fewer polar bears will not save them. Neither will the data collected by scientists like Karyn Rode. Every bear in every zoo could give blood and walk on a treadmill and polar bears would still face a grim future. The only thing that can save polar bears is a sea that stays frozen. And that can happen only if humans stop spewing heat-trapping gases into the sky.

In order for most polar bears to remain in their natural habitat, the Arctic can warm no more than 1.5 degrees Celsius by the end of the century. And people like Gene Agnaboogok and Gilbert Oxereok and Esau Sinnok face similar prospects. The

Native people of Alaska and the polar bears hunt on the same ice under the same sky in the same cold air. They see that the bear Agnaboogok made an orphan and the daughter of that bear, born in a zoo, and the women who raised her and the researchers who studied her and the doctors who mended her broken bones and the people who come to see her are all connected, just as commuters in Columbus and factories in Portland are connected to hunters on the Arctic coast. Just as ice melting in one place causes flooding in another.

After Agnaboogok shot Nora's grandmother, he walked close and shot her again. He did that because his elders taught him that respect means never letting another creature suffer. If you do, they warned him, you will eventually suffer the very same fate. The millions of people who had come to care about Nora knew that at Utah's Hogle Zoo she had, in theory, everything she needed for a good life. She had bucketloads of smelt. She had an expert medical team monitoring her recovery, ready to do whatever it took to make her comfortable. She had a large yard and a pool, and, by the end of 2019, she would again have a polar bear companion to share it with.

The two bears were reunited in December and hardly missed a beat. After a brief bout of open-mouth horseplay to greet each other, their relationship largely picked up where it had left off before Nora's injury.

After Nora went back on exhibit, Jablonski described the bear's condition. "She's not 100 percent, but she's 100 percent of her new normal," the keeper said. Her new normal was the product of malformed joints, anxiety, and a broken humerus. She would never have the body of a fully healthy bear, but she was 100 percent of the bear she could be within the constraints of her various ailments. The term "new normal" gets thrown around a lot with regard to climate change as once-rare extreme weather events become more common. But Nora didn't have a choice about what her new normal would be. We do.

Nora's fans from Ohio will continue to tune in to live broadcasts from the exhibit and flood the comments with well wishes and notes about how much they miss their "Buckeye bear." Visitors to the zoo in Salt Lake City will flock to the windows to watch the bears wrestle and play, the screams of children bouncing off the signs about climate change and shrinking sea ice as the "ambassadors for their species" put a simple face to an infinitely complex problem. There will be elections and, every four or eight years, a new president will occupy the White House. Scientists will continue studying the climate, doing their best to warn the public about what's coming. Those warnings will continue to threaten the profits of monied interests, who will continue waging disinformation campaigns to discredit that science. Polar bears will continue to wander the enclosures of zoos around the world, some pacing anxiously, others seemingly content. They will do so, in theory, to provide a window to a species in peril, so that people who will never travel to the far north can learn to care about something so remote and removed from their everyday lives.

That's the whole point of having polar bears in captivity, as zoos tell it: to give the public a reason to care. Polar bears are classified as "charismatic megafauna" because they are big and charming and people forge an emotional connection with them. Their stories have power, and few animals offer as compelling a story as Nora, an abandoned cub, raised by human hands, who had overcome obstacle after obstacle in her short life. She captured the hearts of thousands of people around the world, people who wanted something to root for.

But what does it say about human beings that we've been unwilling or unable to find those characters to root for within our own species? When the United Nations' 1.5-degree report was released in 2018, headlines splashed across front pages and chyrons scrolled across the bottoms of cable news broadcasts proclaiming we had just twelve years to act. If we didn't reduce emissions by 45 percent in that time, our chances of remaining under the 1.5-degree threshold were small. Pundits and politicians jumped on the 2030 date. Maybe a firm deadline, just over a decade in the future, would provide the urgency to force action.

But the deadline worked both ways. In the world of climate deniers, it was regarded as just another example of Chicken Little alarmism from scientists at the United Nations intent on taking away everyone's cheeseburgers. Even for those who believed the world was warming at an unprecedented rate and humanity was the primary cause, the twelve-year deadline could prove problematic. Twelve years is not a long time for such decisive change to come about, and that kind of deadline could inspire dread and defeatism. If our doom is all but assured, why even try for something that seems so unattainable? But that's not how climate change works.

When the clock strikes midnight on January 1, 2030, the world's oceans will not rise up to swallow the islands of Kiribati, and Shishmaref will not abruptly slide into the sea. Similarly, if we *have* hit the 45 percent target when the sun rises that day, the effects of climate change will not magically go away. Climate change is not pass/fail.

Every action we take will lessen the impacts of supercharged hurricanes and ever-lengthening fire seasons. Every action we fail to take, every week that we kick the can down the road, will increase the likelihood that the ice off the coast of Wales will form later or the next heat wave will set a new record.

Centuries of the Manifest Destiny mindset, in which the natural world was treated as a bottomless resource that could be exploited indefinitely without consequence, have left the world on the verge of unprecedented change, at best. At worst, the planet is in the beginning stages of ecological collapse. Perhaps if we'd heeded the early warning from James Hansen, or if powerful interest groups hadn't spent a fortune undermining science, we could work our way out of this problem by turning down the thermostat or taking public transit to work. That time has passed. Only change on a systemic scale will stave off the worst of what's to come.

Nora's keepers in Columbus kept her warm, held her tight, and waded into the water to show her it was safe. Her handlers in Portland taught her patience. And her keepers in Salt Lake City managed her pain and helped her adapt yet again. But they couldn't conjure the Arctic from concrete or turn the ocean into ice or set Nora free. No matter what anyone did, Nora's life would never unfold as nature intended. But neither would the life of 21736, the bear that Karyn Rode's research team darted on the frozen Chukchi Sea in 2017, on what was probably their last field research trip.

Each one of Nora's caregivers—from Priya Bapodra, the Columbus vet who laid the first of many human hands on her as a cub, to Nicole Nicassio-Hiskey, who oversaw her introduction to Tasul and helped guide her through her most anxious moments, to Joanne Randinitis and Kaleigh Jablonski, who nursed the broken bear back to health—knew the bear's odds were long, and they weighed those odds against the consequences of doing nothing. It took the efforts of dozens of people working thousands of hours to give Nora a chance. Their promise to her was not that they would always get it right, but that they would try.

Acknowledgments

The majority of this book was reported and written on stolen land. If you're in the United States, chances are good that you are reading this on the former territory of Indigenous people who were forced off their land in a sweeping campaign of ethnic cleansing. The city of Portland sits on land historically inhabited by the Multnomah, Wasco, Cowlitz, Kathlamet, Clackamas, Bands of Chinook, Tualatin, Kalapuya, Molalla, and other tribes. Columbus, Ohio, was home to the Shawnee, Miami, Lenape, and Wyandotte peoples. In Salt Lake City, it was the Ute, Diné (Navajo), Paiute, Goshute, and Shoshone tribes. Even in Alaska, where tribes weren't always pushed out of their homes, the legacy of colonialism is active and ongoing. This part of our history is often erased, but to move forward we must acknowledge the past, however horrific it may be.

Getting this project off the ground was no easy feat, and it never would have happened without the faith and support of Mark Katches, editor of *The Oregonian* when the original series was published, in 2017. Kelley Benham French introduced me to the mystifying art of narrative structure. Karly Imus made sure every train got to the station when it should and kept me as close to sane as possible throughout the process. Dave Killen accompanied me on reporting trips and crafted amazing visuals, when he wasn't tripping over himself or almost flying a drone into Russian airspace. My other colleagues at the paper all picked up the slack while I was out chasing a story about a bear cub. To all of you, I am forever grateful.

Over at Neon Literary, my agent, Anna Sproul-Latimer, displayed a bottomless well of patience as she shepherded me through the proposal process. Emma Berry, my editor at Crown Publishing, took a big chance on a first-time author. Thank you to Morgan Baskin, for finding factual errors large and small, and to Doreen Nutaaq Simmonds, who read over the manuscript to look for any inadvertent instances of cultural ignorance.

Each and every one of the zookeepers, veterinarians, curators, and public relations folks at each of the zoos I visited deserves all the gratitude I can offer. Thank you to the scientists who patiently explained their work to me, sometimes two or three times over. I know that I pestered some of you incessantly, and I appreciate your willingness to put up with my never-ending emails promising "just one more follow-up."

I spent a tremendous amount of time over the past several years reporting and writing this book, but there were times when I just couldn't be there and relied on the work of others. I owe a debt of gratitude to the reporters out there who relentlessly pursue the truth and strive to hold the powerful accountable, often with ever-dwindling resources and in the face of assaults on our credibility. If you have the means, please support your local journalists.

On a personal note, I could never have undertaken a project like this without the emotional support of my current and former work spouses, Vivian Ho, Hamed Aleaziz, Lizzy Acker, and Eder Campuzano. You guys are all great.

The biggest thanks of all goes to my wife, Rebecca, who endured so much in the course of this project and who didn't always get the support she needed while I took off on reporting trips for weeks at a time. I love you.

Notes

This book is based on interviews with more than fifty people, conducted over the course of nearly four years; records provided by the zoos; videos of the animals described; and unpublished manuscripts. I also obtained veterinary medical records to describe some ailments in animals and used dozens of academic studies on animal health, the effects of captivity, and climate change.

The scenes at the Columbus Zoo and Aquarium are based on interviews with Priya Bapodra, veterinarian; keepers Devon Sabo and Cindy Cupps; curators Kelly Vineyard and Carrie Pratt; assistant curators Shannon Morarity and Nikki Smith; animal nutrition manager Dana Hatcher; director of visitor engagement Rachel Griffiths; vice president of conservation education Danielle Ross; and vice president of community relations Patty Peters.

I also spoke with Randi Meyerson, former assistant director of animal programs at the Toledo Zoo and coordinator of the Association of Zoos and Aquariums's polar bear Species Survival Plan.

For the scenes in Alaska, I traveled to Wales three times—once in March 2017, then again in March and October of 2019. The scenes in Wales are based on interviews with Gene Rex Agnaboogok, Gilbert Oxereok, Josh Ongtowasruk, Abel Apatiki, Larry Sereadlook, Marie Ningealook, Rachel Seetook, Raymond

Seetook, Clifford Seetook, Terry Crisci, Sean Komonaseak, Dawn Hendrickson, Janelle Cothern, and Joanne Keyes.

In Oregon, I visited the zoo numerous times and observed Nora for countless hours. For the scenes there, I interviewed keepers Nicole Nicassio-Hiskey, Amy Hash, and Jen DeGroot, curator Amy Cutting, veterinarian Mitch Finnegan, and public relations manager Hova Najarian.

For the scenes in Utah, I spoke with keepers Joanne Randinitis and Kaleigh Jablonski, veterinarian Erika Crook, director of conservation Liz Larsen, and community relations coordinator Erica Hansen.

For the scenes detailing Nora's surgery, I spoke at great length with Jeff Watkins, veterinary orthopedic surgeon at Texas A&M University.

For descriptions of polar bear and climate research, I interviewed Karyn Rode and Anthony Pagano about their work both at the Oregon Zoo and in the wild, along with Ian Stirling, who told me about his early research in the Canadian Arctic. Eric Regehr, from the University of Washington's Polar Science Center, helped me understand polar bear population dynamics. His colleagues Axel Schweiger, Harry Stern, and Mike Steele also provided valuable insight into the effects of climate change on ice in the Arctic.

Chapter 1: Abandoned

In addition to the sources cited below, details in this chapter are taken from Nora's weight chart, animal keeper logs, and Aurora's birth plan, all provided by the Columbus Zoo and Aquarium.

4 **When Nora was born:** Columbus Zoo and Aquarium, "Polar Bear Cubs Born at Columbus Zoo," press release, November 12, 2015, https://www.columbuszoo.org/home/about/press-releases/press-release-articles/2015/11/12/polar-bear-cubs-born-at-columbus-zoo.

5   **Most polar bear cubs:** E. Curry, S. Safayi, R. Meyerson, and T. L. Roth, "Reproductive Trends of Captive Polar Bears in North American Zoos: A Historical Analysis," *Journal of Zoo and Aquarium Research* 3, no. 3 (July 2015): 99–106.

7   **It was hot enough:** P. J. Hearty, P. Kindler, H. Cheng, and R. L. Edwards, "A +20 m Middle Pleistocene Sea-Level Highstand (Bermuda and the Bahamas) Due to Partial Collapse of Antarctic Ice," *Geology* 27, no. 4 (April 1, 1999), https://pubs.geoscience world.org/gsa/geology/article-abstract/27/4/375/207100/a-20 -m-middle-pleistocene-sea-level-highstand?redirectedFrom=full text.

7   **The ice sheet that covers Greenland:** A. de Vernal and C. Hillaire-Marcel, "Natural Variability of Greenland Climate, Vegetation, and Ice Volume During the Past Million Years," *Science* 320, no. 5883 (June 20, 2008): 1622–25, https://science.science mag.org/content/320/5883/1622.

7   **At least that's one theory:** S. Liu, E. D. Lorenzen, M. Fumagalli, R. Nielsen, E. Willerslev, J. Wang, et al., "Population Genomics Reveal Recent Speciation and Rapid Evolutionary Adaptation in Polar Bears," *Cell* 157, no. 4 (May 2014): 785–94, https://www .cell.com/cell/fulltext/S0092-8674(14)00488-7#.

7   **Others place the polar bear's divergence:** W. Miller, S. C. Schuster, A. J. Welch, A. Ratan, O. C. Bedoya-Reina, F. Zhao, H. Kim, et al., "Polar and Brown Bear Genomes Reveal Ancient Admixture and Demographic Footprints of Past Climate Change," *Proceedings of the National Academy of Sciences* 109, no. 36 (September 4, 2012), https://www.pnas.org/content/109/36/E2382.

7   **Some say it happened more recently:** J. Stewart, A. M. Lister, I. Barnes, and L. Dalen, "Refugia Revisited: Individualistic Responses of Species in Space and Time," *Proceedings of the Royal Society: Biological Sciences,* October 2009, https:// royalsocietypublishing.org/doi/full/10.1098/rspb.2009.1272.

Chapter 2: A Fateful Hunt

In addition to the sources cited below, Ian Stirling's book *Polar Bears: The Natural History of a Threatened Species* (London: Bloomsbury, 2012) provided valuable background material on polar bear research for this chapter.

13 **Soviet bear counters thought:** Peter Dykstra, "Contentious Fact in Polar Debate Bears Scrutiny," CNN, May 15, 2008, https://scitech.blogs.cnn.com/category/polar-bears/.

13 **In 1973, they formalized that accord:** "Agreement on the Conservation of Polar Bears," Signed in Oslo, Norway, November 15, 1973, https://polarbearagreement.org/resources?task= document.viewdoc&id=1.

13 **The United States banned polar bear hunts:** Sarah Mimms and National Journal, "Why Canada Is Still Stuck with Our Dead Polar Bears," *The Atlantic,* February 4, 2014, https://www .theatlantic.com/politics/archive/2014/02/why-canada-is-still -stuck-with-our-dead-polar-bears/450339/.

13 **Stirling had always been an outdoorsman:** I. Stirling, "Ecology of the Weddell Seal in McMurdo Sound, Antarctica," *Ecology,* July 1, 1969, https://esajournals.onlinelibrary.wiley.com/doi/10 .2307/1936247.

14 **home to one of the world's:** "Polar Bears: A Report About Polar Bears and the State of Resources," Ontario Ministry of Natural Resources, December 2008, accessed April 15, 2020, https:// docs.ontario.ca/documents/3182/state-of-resources-reports -polar-bears.pdf.

14 **Studying polar bears is rarely easy:** Stirling, *Polar Bears,* 66.

14 **Early polar bear research:** Stirling, *Polar Bears,* 68.

15 **If you catch and tag:** "Mark-Recapture," IUCN/SSC Polar Bear Specialist Group, February 26, 2009, accessed April 19, 2020, http://pbsg.npolar.no/en/methods/markrecap.html.

15 **In the southern Beaufort Sea:** "Polar Bear Population Decline a Wake Up Call for Climate Change Action," World Wildlife Fund, accessed April 19, 2020, https://www.worldwildlife.org/ stories/polar-bear-population-decline-a-wake-up-call-for -climate-change-action.

16 **The best guess puts the number:** L. Zuckerman, "Polar Bear Numbers Seen Declining a Third from Arctic Sea Ice Melt," Reuters, December 12, 2016, https://www.reuters.com/article/ us-environment-climate-arctic/polar-bear-numbers-seen-declining -a-third-from-arctic-sea-ice-melt-idUSKBN14205I.

16 **Fossils from a woolly mammoth:** L. Geggel, "Butchered Mam-

moth Suggests Humans Lived in Siberia 45,000 Years Ago," Live Science, January 15, 2016, https://www.livescience.com/53397 -mammoth-human-hunters.html.

17 **Gene Agnaboogok's earliest Inupiat ancestors:** "Thule Culture," University of Alaska Museum of the North, accessed April 19, 2020, https://www.uaf.edu/museum/collections/ archaeo/online-exhibits/paleo-eskimo-cultures/thule/.

Chapter 3: First Feeding

In addition to the sources cited below, details in this chapter come from Aurora's birth plan and Nora's formula recipe, provided by the Columbus Zoo and Aquarium.

19 **They had a manual:** L. Gage, *Hand-Rearing Wild and Domestic Mammals* (Ames: Iowa State Press, 2002).

20 **Fortunately for Hatcher, a veterinarian:** G. Hedberg, A. Derocher, M. Andersen, Q. Rogers, E. Depeters, B. Lönnerdal, L. Mazzaro, R. Chesney, and B. Hollis, "Milk Composition in Free-Ranging Polar Bears (*Ursus maritimus*) as a Model for Captive Rearing Milk Formula," *Zoo Biology* 30, no. 5 (2011): 550–65, https://doi.org/10.1002/zoo.20375.

20 **Hatcher had developed a recipe:** D. Hatcher, "Columbus Zoo and Aquarium's Polar Bear Cub Formula Preparation," internal memo, Columbus Zoo and Aquarium.

22 **The next day, the zoo's public relations:** "Polar Bear Cub Update," Columbus Zoo and Aquarium, YouTube video, 1:16, November 13, 2005, https://www.youtube.com/watch?v=SyutJZyf _1o&t=2s.

23 **In the mid-nineties, a mother bear:** James Brooke, "A Rocky Mountain High: Twin Polar Bears," *New York Times,* December 26, 1995, https://www.nytimes.com/1995/12/26/us/a-rocky -mountain-high-twin-polar-bears.html.

Chapter 4: The Bear

27 **The Upper Mississippi watershed:** S. Changnon, K. Kunkel, and D. Changnon, "Impacts of Recent Climate Anomalies: Los-

ers and Winners," Illinois State Water Survey, case study 2007-01 (June 2007), 17–19, https://www.isws.illinois.edu/pubdoc/DCS/ISWSDCS2007-01.pdf.

28   **He broke his conclusions down:** *Greenhouse Effect and Global Climate Change: Hearing Before the S. Comm. on Energy and Natural Resources,* 100th Cong. (1988) (statement of James Hansen, director, NASA Goddard Inst. for Space Studies), https://babel.hathitrust.org/cgi/pt?id=uc1.b5127807&view=1up&seq=1.

28   **Theophrastus, a student of Aristotle:** "Theophrastus (371–287 BC)," Origins of Botany, Trinity College Dublin, https://www.tcd.ie/Botany/tercentenary/origins/theophrastus.php.

29   **Over the three centuries following:** Douglas W. MacCleery, *American Forests: A History of Resiliency and Recovery* (Durham, NC: Forest History Society, 2012), 21.

29   **Those moving west theorized:** *Encyclopedia of the Great Plains,* s.v. "Rainfall Follows the Plow," accessed April 25, 2020, http://plainshumanities.unl.edu/encyclopedia/doc/egp.ii.049.

29   **Tilled soil released its moisture:** Katie Nodjimbadem, "When the U.S. Government Tried to Make It Rain by Exploding Dynamite in the Sky," SmithsonianMag.com, September 4, 2018, https://www.smithsonianmag.com/history/when-us-government-tried-make-rain-exploding-dynamite-sky-180970193.

29   **A hunter named Jean-Pierre Perraudin:** Holli Riebeek, "Paleoclimatology: Introduction," NASA Earth Observatory, accessed April 25, 2020, https://earthobservatory.nasa.gov/features/Paleoclimatology.

29   **Perraudin brought his idea to:** Encyclopedia.com, "The Discovery of Global Ice Ages by Louis Agassiz," *Science and Its Times: Understanding the Social Significance of Scientific Discovery,* accessed April 25, 2020, https://www.encyclopedia.com/science/encyclopedias-almanacs-transcripts-and-maps/discovery-global-ice-ages-louis-agassiz.

30   **Perhaps the most important question:** Steve Graham, "Svante Arrhenius," NASA Earth Observatory, accessed April 25, 2020, https://earthobservatory.nasa.gov/features/Arrhenius.

30   **Arrhenius's theory, and his research:** Elisabeth Crawford, "Svante Arrhenius," *Encyclopædia Britannica,* February 15, 2020, https://www.britannica.com/biography/Svante-Arrhenius.

31 **Eunice Newton Foote, one of the few:** Amara Huddleston, "Happy 200th Birthday to Eunice Foote, Hidden Climate Science Pioneer," NOAA Climate.gov, July 17, 2019, https://www .climate.gov/news-features/features/happy-200th-birthday -eunice-foote-hidden-climate-science-pioneer.

31 **An entry in the journal *Nature*:** H. A. Phillips, "Pollution of the Atmosphere," *Nature* 27, no. 684 (December 7, 1882), https:// www.nature.com/articles/027127c0.pdf.

31 **Over the following years:** Jeff Nichols (@backwards_river), "1913 Philadelphia Inquirer editorial that questions whether coal and CO2 can account 'for vagaries of the temperature,'" Twitter, October 23, 2016, 3:06 P.M., https://twitter.com/ backwards_river/status/790313357656006656.

31 **and *The Kansas City Star*:** Jeff Nichols (@backwards_river), "Arrhenius got a little play in the American press. Kansas City Star, 1902," Twitter, October 23, 2016, 11:29 A.M., https://twitter .com/backwards_river/status/790258931046035456.

31 **In an 1883 article:** "The Atmosphere," *New York Times,* January 6, 1883, https://timesmachine.nytimes.com/timesmachine/ 1883/01/06/106244132.pdf.

32 **His 1896 study:** S. Arrhenius, "On the Influence of Carbonic Acid in the Air upon the Temperature of the Ground," *Philosophical Magazine and Journal of Science,* series 5, vol. 41 (April 1896): 237–76, https://www.rsc.org/images/Arrhenius1896_ tcm18-173546.pdf.

32 **One of Arrhenius's colleagues, Arvid Högbom:** Patrick Lockerby, "Carbon Cycles by Arvid G. Högbom," Science 2.0, December 21, 2016, https://www.science20.com/the_chatter_box/ blog/carbon_cycles_by_arvid_g_hoegbom-196827.

33 **By the time Hansen went before:** S. R. Weart, "Global Warming, Cold War, and the Evolution of Research Plans," *Historical Studies in the Physical and Biological Sciences* 27, no. 2 (1997): 319–56, https://hsns.ucpress.edu/content/27/2/319.

33 **In 1958, Charles Keeling:** Rob Monroe, "The History of the Keeling Curve," Scripps Institution of Oceanography, April 3, 2013, https://scripps.ucsd.edu/programs/keelingcurve/2013/04/ 03/the-history-of-the-keeling-curve/.

33 **In 1981, Hansen published the first:** J. Hansen et al., "Climate Impact of Increasing Atmospheric Carbon Dioxide," *Science* 213,

no. 4511 (August 28, 1981): 957–66, https://pubs.giss.nasa.gov/abs/ha04600x.html.

34 **Hansen's 1981 paper landed on:** Walter Sullivan, "Study Finds Warming Trend That Could Raise Sea Levels," *New York Times,* August 22, 1981, https://www.nytimes.com/1981/08/22/us/study-finds-warming-trend-that-could-raise-sea-levels.html.

34 **But throughout the early eighties:** C. Sagan et al., "Nuclear Winter: Global Consequences of Multiple Nuclear Explosions," *Science* 222, no. 4630 (December 23, 1983): 1283–92, https://science.sciencemag.org/content/222/4630/1283.

34 **Reagan threatened to cut funding:** S. R. Weart, "Money for Keeling: Monitoring CO_2 Levels," *The Discovery of Global Warming,* American Institute of Physics, July 2008, https://history.aip.org/climate/Kfunds.htm.

34 **A report from the Environmental Protection Agency:** Robert Sangeorge, "EPA Report Predicts Catastrophic Global Warming," UPI, October 18, 1983, https://www.upi.com/Archives/1983/10/18/EPA-report-predicts-catastrophic-global-warming/2626435297600/.

35 **The Reagan administration called the report:** Philip Shabecoff, "Haste of Global Warming Trend Opposed," *New York Times,* October 21, 1983, https://www.nytimes.com/1983/10/21/us/haste-of-global-warming-trend-opposed.html.

35 **Mentions of global warming in:** S. R. Weart, *The Discovery of Global Warming* (Cambridge, MA: Harvard University Press, 2008).

Chapter 5: Signs of Trouble

40 **A small herd of musk oxen:** L. C. Bliss, *Truelove Lowland, Devon Island, Canada: A High Arctic Ecosystem* (Edmonton: University of Alberta Press, 1987).

41 **The Arctic desert landscape:** "Mars Researchers Rendezvous on Remote Arctic Island," NASA Atmospheric Science Data Center, accessed April 26, 2020, https://eosweb.larc.nasa.gov/project/misr/gallery/devon_island.

41 **He and his team set up:** I. Stirling, "Midsummer Observations on Behavior of Wild Polar Bears (Ursus maritimus)," *Canadian*

Journal of Zoology 52, no. 9 (February 12, 1974): 1191–98, https://
www.nrcresearchpress.com/doi/10.1139/z74-157#citart1.

42 **"I wanted to just let the bears":** I. Stirling, "The Amazing
Breeding Behavior of Polar Bears," Polar Bears International,
March 15, 2019, https://polarbearsinternational.org/news/article
-research/the-amazing-breeding-behavior-of-polar-bears/.

42 **Stirling would return to Radstock Bay:** I. Stirling, "Behavior
and Activity Budgets of Wild Breeding Polar Bears (Ursus maritimus)," *Marine Mammal Science* 32, no. 1 (January 2016): 13–37,
https://staging.polarbearsinternational.org/media/3378/2016
-stirling-et-al-breeding-behavior-of-polar-bears.pdf.

46 **Some bears will den as far:** J. W. Lentfer and R. J. Hensel, "Alaskan Polar Bear Denning," *Bears: Their Biology and Management* 4
(1980): 101–8, https://doi.org/10.2307/3872850.

Chapter 6: When Death Came by Dogsled

For descriptions of Wales precontact, I relied heavily on the research
of Ernest S. Burch Jr., specifically his book *Social Life in Northwest
Alaska: The Structure of Iñupiaq Eskimo Nations* (Fairbanks: University of
Alaska Press, 2006). I also owe a debt of gratitude to Tony Hopfinger,
a former reporter at the *Anchorage Daily News,* who wrote an exceptional series called "To Live and Die in Wales," which detailed much
of Wales's history around the time of the Spanish flu. He also helped
me gain access to Henry Greist's unpublished manuscript.

49 **In July of 1776:** "3rd Voyage," Captain James Cook (website),
British Library, accessed April 27, 2020, http://www.captcook-ne
.co.uk/ccne/timeline/voyage3.htm.

49 **But, lured by the promise of:** "Act II: The Third Voyage," *Strait
Through: Magellan to Cook & the Pacific,* Princeton University Library, accessed April 27, 2020, https://lib-dbserver.princeton
.edu/visual_materials/maps/websites/pacific/cook3/cook3
.html.

49 **Oregon's Cape Foulweather was named:** Alwyn Peel, "The
Coast of Oregon," Captain Cook Society, accessed April 27, 2020,
https://www.captaincooksociety.com/home/detail/the-coast-of
-oregon.

50 **"We thought we saw some people":** John Taliaferro, *In a Far Country: The True Story of a Mission, a Marriage, a Murder, and the Remarkable Reindeer Rescue of 1898* (New York: PublicAffairs, 2007), 26.

50 **The village on the cape was:** Burch, *Social Life,* 118.

51 **They hunted bowhead whales:** Burch, *Social Life,* 27.

51 **When marine mammals were scarce:** Burch, *Social Life,* 84.

51 **Most people in Wales lived in:** Burch, *Social Life,* 97.

52 **Every September:** Burch, *Social Life,* 127.

52 **At the center of social life:** Burch, *Social Life,* 119.

52 **Then, around the middle:** Murray Lundberg, "Thar She Blows! Whaling in Alaska and the Yukon," *ExploreNorth* blog, accessed April 27, 2020, http://www.explorenorth.com/library/history/whaling-alaska-yukon.html.

52 **The following year, a rush ensued:** J. Bockstoce, D. Botkin, A. Philp, B. Collins, and J. George, "The Geographic Distribution of Bowhead Whales, Balaena mysticetus, in the Bering, Chukchi, and Beaufort Seas: Evidence from Whaleship Records, 1849–1914," *Marine Fisheries Review* 67, no. 3 (January 2005).

53 **In 1877, the trading brig:** Susan Lebo, "Native Hawaiian Seamen's Accounts of the 1876 Arctic Whaling Disaster and the 1877 Massacre of Alaskan Natives from Cape Prince of Wales," *Hawaiian Journal of History* 40 (2006), https://core.ac.uk/download/pdf/5014805.pdf.

53 **More than a dozen Native men:** Henry Greist, "17 Years with the Eskimo" (unpublished manuscript, 1961), chaps. 2 and 5.

53 **One visitor wrote:** Greist, "17 Years," chaps. 2 and 4.

54 **In 1877, about a decade:** Woman's Home Missionary Society, *19th Annual Report for the Year 1899–1900* (Cincinnati: Western Methodist Book Concern Press, 1900).

54 **Jackson had taught in mission schools:** Richard Dauenhauer, "Two Missions to Alaska," *Pacific Historian* 26 (Spring 1982): 29–41.

54 **To do so, Jackson said:** Taliaferro, *In a Far Country,* 15.

54 **The mission school in Wales:** Taliaferro, *In a Far Country,* 20.

55 **No one knows exactly:** John M. Barry, "How the Horrific 1918 Flu Spread Across America," *Smithsonian,* November 2017, https://www.smithsonianmag.com/history/journal-plague -year-180965222/.

56 **When influenza arrived in Spain:** A. Trilla, G. Trilla, and C. Daer, "The 1918 'Spanish Flu' in Spain," *Clinical Infectious Diseases* 47, no. 5 (September 1, 2008): 668–73, https://academic .oup.com/cid/article/47/5/668/296225.

57 **In August, a new and far deadlier:** Barry, "How the Horrific."

57 **In Columbus, Ohio:** *The American Influenza Epidemic of 1918– 1919: A Digital Encyclopedia,* s.v. "Columbus, Ohio," University of Michigan Center for the History of Medicine and Michigan Publishing, University of Michigan Library, accessed April 27, 2020, https://www.influenzaarchive.org/cities/city-columbus.html#.

57 **The first cases in the Pacific:** *The American Influenza Epidemic of 1918–1919: A Digital Encyclopedia,* s.v. "Seattle, Washington," accessed April 27, 2020, https://www.influenzaarchive.org/cities/ city-seattle.html#.

58 **On October 14, 1918:** S. Mamelund, L. Sattenspiel, and J. Dimka, "1919 Influenza Pandemic in Alaska and Labrador: A Comparison," *Social Science History* 37, no. 2 (2013): 177–229.

59 **When the *Victoria* put in:** Tony Hopfinger, "To Live and Die in Wales, Alaska," *Anchorage Daily News,* November 1, 2007, https:// www.adn.com/rural-alaska/article/live-and-die-wales-alaska/ 2007/11/02/.

60 **A year and a half later:** Greist, "17 Years," chaps. 4 and 2.

61 **One older man:** Greist, "17 Years," chaps. 4 and 3.

61 **Of all the people who died:** Health Analytics and Vital Records, "1918 Pandemic Influenza Mortality in Alaska," Alaska Division of Public Health, accessed April 27, 2020, http://dhss.alaska .gov/dph/VitalStats/Documents/PDFs/AK_1918Flu_DataBrief _092018.pdf.

Chapter 7: Milestones

65 **Brutus and Buckeye:** As it happens, Brutus and Buckeye have a story not unlike Nanuq's: They were orphaned as cubs in Alaska when their mom charged a man out walking his dog. The man

shot the mother in self-defense and helped rescue the brothers. Danielle Barker, "Orphaned Bear Cubs Now Thriving Thanks to Extraordinary Effort of Man Who Shot Their Mom," *USA Today*, August 3, 2018, https://www.usatoday.com/story/news/animal kind/2018/08/03/brother-brown-bears-orphaned-cubs/ 891512002/.

66   **The exhibit had gotten:** "Polar Frontier Opens May 6," press release, Columbus Zoo and Aquarium, May 5, 2010, https:// www.columbuszoo.org/home/about/press-releases/press -release-articles/2010/05/05/polar-frontier-opens-may-6.

67   **In the 1930s, Austrian scientist:** Saul McLeod, "Konrad Lorenz's Imprinting Theory," Simply Psychology, 2018, https:// www.simplypsychology.org/Konrad-Lorenz.html.

67   **"The man who walked with geese":** "Konrad Lorenz—Facts," NobelPrize.org, Nobel Media AB 2020, accessed April 27, 2020, https://www.nobelprize.org/prizes/medicine/1973/lorenz/ facts/.

67   **At the Monterey Bay Aquarium:** Katia Hetter, "These Sea Otters Adopt Orphaned Pups and Raise Them to Be Wild," CNN Travel, September 25, 2019, https://www.cnn.com/travel/article/ sea-otters-monterey-bay-aquarium-california/index.html.

68   **Rulers in ancient Egypt:** Mark Rose, "World's First Zoo— Hierakonpolis, Egypt," *Archaeology* 63, no. 10 (January/February 2010), https://archive.archaeology.org/1001/topten/egypt.html.

68   **Roman emperors kept hundreds of animals:** Caroline Wazer, "The Exotic Animal Traffickers of Ancient Rome," *The Atlantic*, March 31, 2016, https://www.theatlantic.com/science/archive/ 2016/03/exotic-animals-ancient-rome/475704/.

69   **King Henry III kept one:** "The Tower of London Menagerie," Historic Royal Palaces, accessed April 27, 2020, https://www.hrp .org.uk/tower-of-london/history-and-stories/the-tower-of -london-menagerie/#gs.4odzuw.

69   **Five hundred years later:** Michael Engelhard, "Furry Attractions: Polar Bears in the Zoo," Center for Humans & Nature, accessed April 27, 2020, https://www.humansandnature.org/ furry-attractions-polar-bears-in-the-zoo.

69   **In the early 1800s:** "Landmarks in ZSL History," Zoological Society of London, accessed April 28, 2020, https://www.zsl.org/ about-us/landmarks-in-zsl-history.

70 **As one menagerie owner put it:** Michael Engelhard, "Polar Attraction: A Brief History of the Arctic White Bear in Captivity," *Journal of Wild Culture,* February 26, 2017, https://www .wildculture.com/article/polar-attraction-brief-history-arctic -white-bear-captivity/1624.

71 **Of the 2,800 licensed animal exhibitors:** "About AZA Accreditation," Association of Zoos and Aquariums, accessed April 28, 2020, https://www.aza.org/what-is-accreditation.

72 **Nora's debut was covered:** 10TV Web Staff, "Polar Bear Cub Nora Makes Debut at Columbus Zoo," WBNS-10TV Columbus, Ohio, May 17, 2016, https://www.10tv.com/article/polar-bear -cub-nora-makes-debut-columbus-zoo.

Chapter 8: Farewell

In addition to the sources cited below, *Ice Bear: The Cultural History of an Arctic Icon* (Seattle: University of Washington Press, 2017), by Michael Engelhard, provided essential details not just about Knut but about the role of the polar bear as a symbol throughout history.

74 **There was Heidi the cross-eyed:** Mary Beth Warner, "A Star Is Born: Heidi, the Cross-Eyed Opossum, Charms Germany," *Der Spiegel,* January 7, 2011, https://www.spiegel.de/international/ zeitgeist/a-star-is-born-heidi-the-cross-eyed-opossum-charms -germany-a-738309.html.

74 **There was Koko:** Douglas Main, "Why Koko the Gorilla Mattered," *National Geographic,* June 21, 2018, https://www.national geographic.com/news/2018/06/gorillas-koko-sign-language -culture-animals/.

74 **Harriet, a Galápagos tortoise:** Associated Press, "176-Year-Old 'Darwin's Tortoise' Dies in Zoo," NBC News, June 24, 2006, http://www.nbcnews.com/id/13115101/ns/world_news-asia _pacific/t/-year-old-darwins-tortoise-dies-zoo/#.Xqm VIZNKi7M.

75 **Born in December of 2006:** Kate Connolly, "Rejected at Birth, Knut Becomes Berlin Zoo's Bear Essential," *Guardian,* March 24, 2007, https://www.theguardian.com/world/2007/mar/24/animal welfare.germany.

75 **He was raised by a keeper:** Engelhard, *Ice Bear,* 21–25.

75 **"Each time his keeper leaves him":** "Berlin Rallies Behind Baby Bear," BBC News, March 20, 2007, http://news.bbc.co.uk/1/hi/world/europe/6470509.stm.

76 **His public debut:** Michael Engelhard, "How One Polar Bear Cub Had Germany (and the World) Enthralled," *National Geographic*, October 13, 2016, https://blog.nationalgeographic.org/2016/10/13/how-one-polar-bear-cub-had-germany-and-the-world-enthralled/.

76 **Police were called in to protect:** Kate Connolly, "Guards Protect Knut After Death Threat," *Guardian*, April 19, 2007, https://www.theguardian.com/world/2007/apr/20/germany.kateconnolly.

76 **But even as some soured:** "End of an Era Nearing: Knut Steadily Getting Less Cute," *Der Spiegel*, April 30, 2007, https://www.spiegel.de/international/zeitgeist/end-of-an-era-nearing-knut-steadily-getting-less-cute-a-480321.html.

77 **A protracted legal battle played out:** "Star Bear Knut to Stay in Berlin," BBC News, July 8, 2009, http://news.bbc.co.uk/2/hi/europe/8140185.stm.

77 **He needed his independence, too:** Associated Press, "Berlin Zoo Makes Polar Bear Cub a Solo Act," CBS News, July 12, 2007, https://www.cbsnews.com/news/berlin-zoo-makes-polar-bear-cub-a-solo-act/.

77 **"He doesn't know that he's":** "Polar Bear Missing Human Contact: Knut Pining for His Lost Friends," *Der Spiegel*, March 25, 2008, https://www.spiegel.de/international/zeitgeist/polar-bear-missing-human-contact-knut-pining-for-his-lost-friends-a-543145.html.

77 **In September of 2008:** David Crossland, "Berlin Polar Bear Alone at Home: Knut Relieved at Departure of Italian Girlfriend," *Der Spiegel*, August 2, 2010, https://www.spiegel.de/international/zeitgeist/berlin-polar-bear-alone-at-home-knut-relieved-at-departure-of-italian-girlfriend-a-709771.html.

77 **He spent most of his time:** "Knut the Polar Bear Bullied," CBS News, YouTube video, 0:29, October 10, 2010, https://www.youtube.com/watch?v=cICyHji6JYE.

78 **"For the time being":** Claire McCormack, "Knut the Polar Bear: Unlucky in Love," *Time*, October 20, 2010, https://newsfeed.time.com/2010/10/20/knut-the-polar-bear-unlucky-in-love/.

78 **On March 19, 2011:** "Knut the Polar Bear Dies," YouTube video, 0:51, uploaded by melonheadman, March 22, 2011, https://www.youtube.com/watch?v=OMZXu303tms.

78 **"He had a special place":** Associated Press, "Knut, the Polar Bear Raised by Berlin Zoo Keepers, Dies in Compound," *Guardian,* March 19, 2011, https://www.theguardian.com/world/2011/mar/19/knut-polar-bear-berlin-dies.

78 **Knut himself was taxidermied:** Allison Meier, "Immortalized in Taxidermy, Berlin's Favorite Polar Bear Knut Returns," *Atlas Obscura,* March 27, 2013, https://www.atlasobscura.com/articles/knut-the-polar-bear-taxidermy-on-display.

78 **Thousands signed an online memorial book:** Mary Beth Warner, "Remembering Knut: Polar Bear Will Be Given Home at Natural History Museum," *Der Spiegel,* March 25, 2011, https://www.spiegel.de/international/germany/remembering-knut-polar-bear-will-be-given-home-at-natural-history-museum-a-753142.html.

79 **Gerald Uhlich, a trustee:** "Polar Bear Turned Cash Cow: Knut the Business-Bear," *Der Spiegel,* May 11, 2007, https://www.spiegel.de/international/zeitgeist/polar-bear-turned-cash-cow-knut-the-business-bear-a-482368.html.

79 **"There's the abstract and the theory":** "U.S. Special Representative for the Arctic and Norwegian Ambassador Visit Columbus Zoo and Aquarium," YouTube video, 2:25, Columbus Zoo and Aquarium, July 15, 2016, https://www.youtube.com/watch?v=O_IAqrQ38ew.

80 **In March of 2016:** Natasha Vizcarra, "Another Record Low for Arctic Sea Ice Maximum Winter Extent," National Snow & Ice Data Center, March 26, 2016, https://nsidc.org/arcticseaice news/2016/03/another-record-low-for-arctic-sea-ice-maximum-winter-extent/.

81 **Just a few weeks before Nora:** "Fact Sheet: What Climate Change Means for Your Health and Family," National Archives and Records Administration, accessed May 8, 2020, https://obamawhitehouse.archives.gov/the-press-office/2016/04/04/fact-sheet-what-climate-change-means-your-health-and-family.

82 **It set the stage for:** *Agenda 21,* report of the United Nations Conference on Environment and Development, June 3–14, 1992,

https://sustainabledevelopment.un.org/content/documents/
Agenda21.pdf.

82 **Its lobbyists met with UN scientists:** Climate Investigations
Center, "1996 GCC STAC October Meeting Minutes," Docu-
mentCloud, September 24–26, 1996, accessed May 8, 2020,
https://www.documentcloud.org/documents/5631463-AIAM
-051494.html#document/p91/a494375.

82 **The coalition wrote that:** Andrew Revkin, "Industry Ignored Its
Scientists on Climate," *New York Times,* April 24, 2009, https://
www.nytimes.com/2009/04/24/science/earth/24deny.html.

83 **After the Earth Summit agreement:** John Vidal, "Revealed:
How Oil Giant Influenced Bush," *Guardian,* June 8, 2005, https://
www.theguardian.com/news/2005/jun/08/usnews.climate
change.

83 **By the time the UN released:** IPCC, *Fourth Assessment Report,*
2007, accessed May 8, 2020, https://www.ipcc.ch/assessment
-report/ar4/.

Chapter 9: Tasul

As part of the original series on which this book is based, published
by *The Oregonian*/OregonLive in 2017, Grant Butler did a great deal of
research on the history of the Oregon Zoo. His work was instrumental in the writing of this chapter.

87 **They can identify other individuals:** "Bear Intelligence," *Nature,* PBS, June 10, 2008, http://www.pbs.org/wnet/nature/
arctic-bears-bear-intelligence/779/.

87 **Studies of sun bears:** Meilan Solly, "Sun Bears Mimic Each
Other's Facial Expressions to Communicate," *Smithsonian,*
March 22, 2019, https://www.smithsonianmag.com/smart
-news/sun-bears-mimic-each-others-facial-expressions
-communicate-180971778/.

91 **He was hoping to sell:** "Letter from Portland Apothecary Richard Knight, Dated June 6, 1888, Regarding a Male Brown Bear
and She Grizzly He Brought to the Area and Wanted to Sell,"
Oregon Zoo, accessed May 9, 2020, https://www.oregonzoo.org/
gallery/images/letter-portland-apothecary-richard-knight-dated
-june-6-1888-regarding-male-brown-bear.

91 **He offered to donate the grizzly:** "History," Oregon Zoo, accessed May 9, 2020, https://www.oregonzoo.org/about/about -oregon-zoo/history.

92 **"I am of the opinion":** Grant Butler, "Polar Bears at Portland's Zoo: A Century Marked by Sadness, Tragedy," *The Oregonian/OregonLive*, October 18, 2017, https://www.oregonlive.com/ portland/2017/10/polar_bears_at_portlands_zoo_a.html.

96 **Tasul's monumental blood draw:** Katy Muldoon, "Oregon Zoo Makes a Medical Breakthrough with Polar Bears," *The Oregonian/OregonLive*, September 19, 2012,, https://www.oregonlive.com/ portland/2012/09/oregon_zoo_makes_a_medical_bre.html.

97 **The researchers also strapped a GoPro:** "Tasul's Collar," YouTube video, 2:44, Oregon Zoo, July 30, 2013, https://www .youtube.com/watch?v=4qEbPLrVOb0&feature=youtu.be.

97 **Over the course of three years:** Robinson Meyer, "What Scientists Learned from Strapping a Camera to a Polar Bear," *The Atlantic*, February 2, 2018, https://www.theatlantic.com/science/ archive/2018/02/what-scientists-learned-from-strapping-a -camera-to-a-polar-bear/552083/.

Chapter 11: Arrival

In addition to the sources cited below, *Animal Madness: Inside Their Minds* (New York: Simon & Schuster, 2015), by Laurel Braitman, proved an invaluable resource on the use of pharmaceuticals in zoo animals and the subject of intellect in the animal world.

Documents: Nora's medical charts, courtesy of the Oregon Zoo.

110 **In the news release:** "Nora Update: Young Bear Meets Elderly Companion Tasul," Oregon Zoo, accessed May 9, 2020, https:// www.oregonzoo.org/node/3194/media.

111 **A 2003 study from:** R. Clubb and G. Mason, "Captivity Effects on Wide-Ranging Carnivores," *Nature* 425, no. 473–74 (October 2, 2003), https://www.nature.com/articles/425473a.

112 **Michael Hutchins, then director:** Mark Derr, "Big Beasts, Tight Space and a Call for Change," *New York Times*, October 2, 2003, https://www.nytimes.com/2003/10/02/us/big-beasts-tight -space-and-a-call-for-change.html.

113 **Regular meals and consistent medical care:** M. Tidière, J. Gaillard, V. Berger, D. W. H. Müller, L. B. Lackey, O. Gimenez, M. Clauss, and J. Lemaître, "Comparative Analyses of Longevity and Senescence Reveal Variable Survival Benefits of Living in Zoos Across Mammals," *Scientific Reports* 6, no. 36361 (November 7, 2016), https://www.nature.com/articles/srep36361.

113 **About ninety miles southeast of Columbus:** "The Wilds 2019 Media Kit," Columbus Zoo and Aquarium, accessed May 9, 2020, https://columbuszoo.org/docs/default-source/media-tools -content/the-wilds-media-kit-2019.pdf?sfvrsn=24e6b4a6_2.

114 **in Columbus, one of Nora's keepers:** "Wildlife and Wild Places—Polar Bear Cub Update," YouTube video, 3:00, Columbus Zoo and Aquarium, December 31, 2015, https://www .youtube.com/watch?v=hMpTPtig9uw.

115 **Born to a polar bear:** Blade Staff and News Services, "Popular Polar Bear Born in Toledo Dies," *Blade* (Toledo, OH), August 29, 2013, https://www.toledoblade.com/news/2013/08/30/Popular -polar-bearborn-in-Toledo-dies/stories/20130829212.

115 **Fresh off a $35 million renovation:** "At Last, a Joy for All Ages: Central Park Zoo Is Back," *New York Times,* August 9, 1988, https://timesmachine.nytimes.com/timesmachine/1988/08/09/ 495388.html?pageNumber=1.

115 **the zoo boasted a snow monkey island:** "Japanese Macaque," Central Park Zoo, October 17, 2017, https://www.centralpark .com/things-to-do/central-park-zoo/japanese-macaque/.

115 **It had a revamped tropical zone:** "Tropic Zone: The Rainforest," Central Park Zoo, accessed May 9, 2020, https:// centralparkzoo.com/exhibits/tropic-zone-the-rainforest.

115 **Sometime in the early to mid-nineties:** "Endless Pools & Gus, the Central Park Zoo Polar Bear," YouTube video, 2:35, Endless Pools, May 25, 2011, https://www.youtube.com/watch?v= RMy3wt0hDxk.

116 **It's not unique to animals:** N. R. Kleinfield, "Farewell to Gus, Whose Issues Made Him a Stat," *New York Times,* August 28, 2013, https://www.nytimes.com/2013/08/29/nyregion/gus-new -yorks-most-famous-polar-bear-dies-at-27.html.

116 **In the 1600s, René Descartes:** "Rene Descartes," How to Do Animal Rights, accessed May 9, 2020, http://www.animalethics .org.uk/descartes.html.

116 **About a hundred years later:** *Internet Encyclopedia of Philosophy,*
s.v. "Animal Minds," accessed May 9, 2020, https://www.iep.utm
.edu/ani-mind/.

116 **William Lauder Lindsay, a nineteenth-century:** Edmund
Ramsden and Duncan Wilson, "The Suicidal Animal: Science
and the Nature of Self-Destruction," *Past & Present* 224, no. 1
(August 2014), https://academic.oup.com/past/article/224/1/
201/1411207.

116 **Charles Darwin shared similar views:** Thomas Seeley and Paul
Sherman, "History and Basic Concepts," in *Encyclopædia Britannica,* accessed May 9, 2020, https://www.britannica.com/science/
animal-behavior/History-and-basic-concepts.

117 **In 1970, Gordon Gallup Jr.:** G. Gallup Jr., "Chimpanzees: Self-
Recognition," *Science* 167, no. 3914 (January 2, 1970): 86–87.

117 **The mirror test:** "List of Animals That Have Passed the Mirror
Test," Animal Cognition, October 29, 2016, http://www.animal
cognition.org/2015/04/15/list-of-animals-that-have-passed-the
-mirror-test/.

117 **Though the mirror test has been:** Ed Yong, "Can Dogs Smell
Their 'Reflections'?," *The Atlantic,* August 18, 2017, https://www
.theatlantic.com/science/archive/2017/08/can-dogs-smell-their
-reflections/537219/.

117 **The Cambridge Declaration on Consciousness:** Philip Low,
"The Cambridge Declaration on Consciousness," signed at the
Francis Crick Memorial Conference on Consciousness in
Human and Non-Human Animals, at Churchill College, Univer-
sity of Cambridge, July 7, 2012, http://fcmconference.org/img
/CambridgeDeclarationOnConsciousness.pdf.

118 **As scientists sought to understand:** "Affective States," Animal
Behavior and Cognition Lab, Department of Animal Science,
UC Davis, accessed May 9, 2020, https://horback.faculty.ucdavis
.edu/assessing-affective-states/.

118 **In her book *Animal Madness:*** Braitman, *Animal Madness,* 109.

119 **One of the first was a gorilla:** Braitman, *Animal Madness,* 197–
204.

120 **"[Psychotropic drugs are] definitely a wonderful":** Jenni
Laidman, "Zoo Using Drugs to Help Manage Anxious Animals,"
Blade (Toledo, OH), September 12, 2005, https://www.toledo

blade.com/frontpage/2005/09/12/Zoos-using-drugs-to-help -manage-anxious-animals.html.

120 **The market for psychotropic drugs:** Braitman, *Animal Madness,* 195.

121 **"He liked to see them scream":** Tad Friend, "It's a Jungle in Here," *New York,* April 24, 1995, https://books.google.com/.

121 **The Canadian band:** Braitman, *Animal Madness,* 210.

121 **A *New York Times* columnist responded:** John Kifner, "About New York: Stay-at-Home SWB, 8, Into Fitness, Seeks Thrills," *New York Times,* July 2, 1994, https://timesmachine.nytimes .com/timesmachine/1994/07/02/issue.html.

122 **Gus's enclosure, which he shared:** Braitman, *Animal Madness,* 211.

Chapter 12: Sinking into the Sea

To describe Karyn Rode's fieldwork, I relied on a great video called "Breakthrough: Polar Bear Witness," directed by Luke Groskin of *Science Friday,* who traveled to the Chukchi Sea in 2017.

123 **Off the coast of Alaska:** "Breakthrough: Polar Bear Witness," directed by Luke Groskin, produced by Emily Driscoll and Luke Groskin, *Science Friday,* June 29, 2017, https://www .sciencefriday.com/videos/polar-bear-witness/.

124 **In 2005, the Center for Biological:** Kassie Siegel and Brendan Cummings, "Petition to List the Polar Bear as a Threatened Species Under the Endangered Species Act," Center for Biological Diversity, February 16, 2005, https://www.biologicaldiversity.org/ species/mammals/polar_bear/pdfs/15976_7338.pdf.

125 **The Center usually petitioned on behalf:** "Spotted Owl Action Timeline," Center for Biological Diversity, accessed May 9, 2020, https://www.biologicaldiversity.org/species/birds/northern _spotted_owl/action_timeline.html.

125 **Over the ensuing year:** "Future Retreat of Arctic Sea Ice Will Lower Polar Bear Populations and Limit Their Distribution," U.S. Geological Survey, September 7, 2007, https://archive.usgs .gov/archive/sites/www.usgs.gov/newsroom/article.asp-ID= 1773.html.

125 **The public was invited to weigh in:** "Endangered and Threatened Wildlife and Plants; Determination of Threatened Status for the Polar Bear (Ursus maritimus) Throughout Its Range; Final Rule," Department of the Interior, May 15, 2008, https://www.fws.gov/r7/fisheries/mmm/polarbear/pdf/Polar_Bear_Final_Rule.pdf.

126 **Political reactions fell along:** Larry Greenemeier, "U.S. Protects Polar Bears Under Endangered Species Act," *Scientific American,* May 14, 2008, https://www.scientificamerican.com/article/polar-bears-threatened/.

126 **The Endangered Species Act:** Juliet Eilperin, "Polar Bear Is Named 'Threatened,'" *Washington Post,* May 15, 2008, https://www.washingtonpost.com/wp-dyn/content/article/2008/05/14/AR2008051401596.html.

126 **But listing the bears as threatened:** "Polar Bear Critical Habitat Questions & Answers," U.S. Fish and Wildlife Service, accessed May 9, 2020, https://www.fws.gov/r7/fisheries/mmm/polarbear/pdf/Updated%20FAQs%20for%20polar%20bear%20CH.pdf?SiteName=FWS&Entity=PRAsset&SF_PRAsset_PRAssetID_EQ=131878&XSL=PressRelease&Cache=True.

127 **While more progressive on environmental issues:** Michael Stickings, "Too Bad for Those Threatened Polar Bears," RealClearPolitics, May 15, 2008, https://www.realclearpolitics.com/cross_tabs/2008/05/too_bad_for_those_threatened_p.html.

128 **In 2019, Eric Regehr:** E. Regehr, N. Hostetter, R. Wilson, K. Rode, M. St. Martin, and S. Converse, "Integrated Population Modeling Provides the First Empirical Estimates of Vital Rates and Abundance for Polar Bears in the Chukchi Sea," *Scientific Reports* 8, no. 16780 (November 14, 2018), https://www.nature.com/articles/s41598-018-34824-7.

128 **Summer sea ice in the Chukchi:** "Polar Bear Status Table," IUCN/SSC Polar Bear Specialist Group, September 2019, accessed May 9, 2020, http://pbsg.npolar.no/export/sites/pbsg/en/docs/2019-PBSG-StatusTable.pdf.

129 **Between 2000 and 2010, researchers saw:** "Southern Beaufort Sea Polar Bear Population Declined in the 2000s," U.S. Geological Survey, November 17, 2014, https://www.usgs.gov/news/southern-beaufort-sea-polar-bear-population-declined-2000s.

Chapter 13: Alone Again

In addition to the sources cited below, details in this chapter come from Tasul's medical charts, courtesy of the Oregon Zoo.

133 **he called it a "hoax":** Donald J. Trump (@realDonaldTrump), "Snowing in Texas and Louisiana, record setting freezing temperatures throughout the country and beyond. Global warming is an expensive hoax!," Twitter, January 28, 2014, 11:27 P.M., https://twitter.com/realDonaldTrump/status/428414113463 955457.

133 **"bullshit":** Donald J. Trump (@realDonaldTrump), "This very expensive GLOBAL WARMING bullshit has got to stop. Our planet is freezing, record low temps, and our GW scientists are stuck in ice," Twitter, January 1, 2014, 5:39 P.M., https://twitter.com/realDonaldTrump/status/418542137899491328.

133 **"created by and for the Chinese":** Donald J. Trump (@real DonaldTrump), "The concept of global warming was created by and for the Chinese in order to make U.S. manufacturing non-competitive," Twitter, November 6, 2012, 2:15 P.M., https://twitter.com/realDonaldTrump/status/265895292191248385.

133 **In January of 2016:** "2015 Was the Hottest Year on Record," NASA Earth Observatory, accessed May 9, 2020, https://earthobservatory.nasa.gov/images/87359/2015-was-the-hottest-year-on-record.

133 **"Who's going to protect the environment":** Christopher Solomon, "The Donald Trump Environmental Scorecard," *Outside,* September 26, 2016, https://www.outsideonline.com/2117796/donald-trump-environmental-scorecard.

134 **Soon after his election:** S. R. Weart, "Money for Keeling: Monitoring CO_2 Levels," *The Discovery of Global Warming,* American Institute of Physics, July 2008, https://history.aip.org/climate/Kfunds.htm.

Chapter 14: The Last Skin Boat

139 **According to the most recent estimates:** "Selected Characteristics of the Total and Native Populations in the United States: Wales, Alaska," American Community Survey 2017, U.S. Census Bureau, https://data.census.gov/cedsci/table?q=Wales,

%20Alaska&g=1600000US0282860&hidePreview=false&tid
=ACSST5Y2018.S0601&vintage=2018&layer=VT_2018_160_00
_PY_D1&cid=DP05_0001E.

139 **twice the unemployment rate for Alaska:** "Seasonally Adjusted
Unemployment Rates Alaska and the U.S., January 2012 to April
2020," Alaska Department of Labor and Workforce Development, accessed May 9, 2020, https://live.laborstats.alaska.gov/
labforce/.

141 **He read in** *National Geographic:* Marianne Lavelle, "Arctic Shipping Soars, Led by Russia and Lured by Energy," *National Geographic,* May 22, 2016, https://www.nationalgeographic.com/
news/energy/2013/11/131129-arctic-shipping-soars-led-by
-russia/.

141 **Using data from radio collars:** G. Durner, D. Douglas, S. Al-
beke, J. Whiteman, S. Amstrup, E. Richardson, R. Wilson, and M.
Ben-David, "Increased Arctic Sea Ice Drift Alters Adult Female
Polar Bear Movements and Energetics," *Global Change Biology* 23,
no. 9 (2017), https://onlinelibrary.wiley.com/doi/abs/10.1111/
gcb.13746.

142 **In 2014, the Inuit Circumpolar Council:** "About ICC," Inuit
Circumpolar Council, accessed May 9, 2020, https://www
.inuitcircumpolar.com/about-icc/.

142 **They tapped local experts:** "Bering Strait Regional Food Security Workshop: How to Assess Food Security from an Inuit Perspective: Building a Conceptual Framework on How to Assess
Food Security in the Alaskan Arctic," Inuit Circumpolar
Council–Alaska, April 14–15, 2014, https://iccalaska.org/wp
-icc/wp-content/uploads/2016/03/Bering-Strait-Regional-WS
-Repor_0319.pdf.

143 **An economic disaster was declared:** Suzanna Caldwell, "Disaster Declared for Subsistence Walrus Hunt on St. Lawrence Island," *Anchorage Daily News,* September 28, 2016, https://www
.adn.com/rural-alaska/article/disaster-declared-subsistence
-walrus-hunt-st-lawrence-island/2013/09/03/.

Chapter 15: Another Hurdle

In addition to the sources cited below, details in this chapter come
from Tasul's medical charts, courtesy of the Oregon Zoo.

148 **It had a massive pool:** Tiffany Frandsen, "Hogle Zoo to Put Polar Bear Rizzo on End-of-Life Care," *Salt Lake Tribune*, August 10, 2017, https://www.sltrib.com/news/2017/04/10/hogle -zoo-to-put-polar-bear-rizzo-on-end-of-life-care/.

148 **Hogle also had a strong conservation record:** "Our Big Six," Utah's Hogle Zoo, accessed May 9, 2020, https://www.hoglezoo .org/meet_our_animals/conservation/carousel-for-conservation -projects/.

149 **In August, Joanne Randinitis:** "Live from Oregon Zoo with Nora the Polar Bear and Hogle Zoo Keeper, Joanne!," Facebook Live video, 12:20, Utah's Hogle Zoo, August 4, 2017, https:// www.facebook.com/HogleZoo/videos/10159145897340173/ ?comment_id=10159145970010173&__tn__=R.

149 **In June, a stretch of three:** "Portland 2017 Weather Recap," National Weather Service, December 31, 2017, https://www.weather .gov/pqr/2017recappdx.

150 **That's where Liz FitzGerald was headed:** KGW, " 'Do You Realize You Just Started a Forest Fire?': Witness to Teen Suspect," KGW News, September 6, 2017, https://www.kgw.com/article/ news/do-you-realize-you-just-started-a-forest-fire-witness-to -teen-suspect/283-471324692.

151 **The officer quickly called it in:** "Timeline of the Eagle Creek Fire," U.S. Department of Agriculture, U.S. Forest Service, accessed May 9, 2020, https://usfs.maps.arcgis.com/apps/Cascade/ index.html?appid=d2a772ec2fb14555a51a0ac0a08a1116.

153 **Hardware stores ran low on masks:** Andrew Theen, "Eagle Creek Fire: Smoke-Socked Portland Could See Reprieve as Winds Shift," *The Oregonian*/OregonLive, September 6, 2017, https://www.oregonlive.com/wildfires/2017/09/eagle_creek_fire _smoke-socked.html.

153 **On September 5 alone:** "Impacts of Oregon's 2017 Wildfire Season," Oregon Forest Resources Institute, January 2, 2018, https://oregonforests.org/sites/default/files/2018-01/OFRI %202017%20Wildfire%20Report%20-%20FINAL%2001-02-18 .pdf.

153 **Four homes burned down:** Jim Ryan, "By the Numbers: A Look Back at the Eagle Creek Fire, 3 Months Later," *The Oregonian*/

OregonLive, December 8, 2017, https://www.oregonlive.com/wildfires/2017/12/by_the_numbers_a_look_back_at.html.

153 **The fifteen-year-old boy:** Maxine Bernstein, "Teen Admits Starting Eagle Creek Fire, Sentenced to 5 Years of Probation," *The Oregonian*/OregonLive, February 16, 2018, https://www.oregonlive.com/wildfires/2018/02/teen_accused_of_setting_eagle.html.

154 **A week before the fire:** Eric Blake and David Zelinsky, "Hurricane Harvey," National Hurricane Center Tropical Cyclone Report, January 23, 2018, https://www.nhc.noaa.gov/data/tcr/AL092017_Harvey.pdf.

154 **Though the recorded death toll:** John D. Sutter and Leyla Santiago, "Hurricane Maria Death Toll May Be More Than 4,600 in Puerto Rico," CNN, May 29, 2018, https://www.cnn.com/2018/05/29/us/puerto-rico-hurricane-maria-death-toll/index.html.

154 **Dozens were killed:** Alex Emslie, "October Fires' 44th Victim: A Creative, Globetrotting Engineer with 'The Kindest Heart,'" KQED, November 28, 2017, https://www.kqed.org/news/11633757/october-fires-44th-victim-a-creative-globetrotting-engineer-with-the-kindest-heart.

154 **"The gorge isn't just a recreation":** Jamie Hale, "As Eagle Creek Fire Rages, Why We Mourn for the Gorge," *The Oregonian*/OregonLive, September 6, 2017, https://www.oregonlive.com/wildfires/2017/09/as_eagle_creek_fire_rages_why.html.

155 **In the case of Harvey, flooding:** W. Zhang, G. Villarini, G. Vecchi, and J. Smith, "Urbanization Exacerbated the Rainfall and Flooding Caused by Hurricane Harvey in Houston," *Nature* 563 (November 14, 2018), https://www.nature.com/articles/s41586-018-0676-z.

155 **In Puerto Rico, mismanagement:** Alexia Fernández Campbell, "It Took 11 Months to Restore Power to Puerto Rico After Hurricane Maria. A Similar Crisis Could Happen Again," *Vox*, August 15, 2018, https://www.vox.com/identities/2018/8/15/17692414/puerto-rico-power-electricity-restored-hurricane-maria.

155 **The resulting death toll:** J. Robine, S. Cheung, S. Le Roy, H. Van Oyen, C. Griffiths, J. Michel, and F. Herrmann, "Death Toll Ex-

ceeded 70,000 in Europe During the Summer of 2003," *Comptes Rendus Biologies* 331, no. 2 (February 2008): 171–78, https://www.sciencedirect.com/science/article/pii/S1631069107003770 ?via=ihub.

156 **One group of scientists:** P. Stott, D. A. Stone, and M. R. Allen, "Human Contribution to the European Heatwave of 2003," *Nature* 432 (December 2, 2004), https://www.nature.com/articles/nature03089.

156 **In 2016, a team of researchers:** D. Mitchell, C. Heaviside, S. Vardoulakis, C. Huntingford, G. Masato, B. Guillod, P. Frumhoff, A. Bowery, D. Wallom, and M. Allen, "Attributing Human Mortality During Extreme Heat Waves to Anthropogenic Climate Change," *Environmental Research Letters* 11, no. 7 (July 8, 2016), https://iopscience.iop.org/article/10.1088/1748-9326/11/7/074006.

156 **Using climate models and historical drought:** J. Abatzoglou and A. Williams, "Impact of Anthropogenic Climate Change on Wildfire Across Western US Forests," *Proceedings of the National Academy of Sciences* 113, no. 42 (October 10, 2016): 11770–75, https://www.pnas.org/content/113/42/11770.

Chapter 16: On the Edge of a Warming World

159 **The highest point on the island:** "Shishmaref Site Analysis for Potential Emergency Evacuation and Permanent Relocation Sites," National Resource Conservation Service and Shishmaref Erosion and Relocation Coalition, accessed May 10, 2020, https://www.commerce.alaska.gov/web/portals/4/pub/shishmaref_relocation_site_reconnaissance_nrcs.pdf.

160 **Over the course of just a few:** "Shishmaref Relocation Strategic Plan," Shishmaref Erosion and Relocation Coalition, January 2002, https://www.cakex.org/sites/default/files/documents/strategic_plan_final_200211.pdf.

160 **None of the strategies provided:** O. Mason, J. Jordan, L. Lestak, and W. Manley, "Narratives of Shoreline Erosion and Protection at Shishmaref, Alaska: The Anecdotal and the Analytical," in *Pitfalls of Shoreline Stabilization: Selected Case Studies*, ed. J. Andrew G. Cooper and Orrin H. Pilkey (Heidelberg: Springer, Dordrecht, 2012).

160   **More than a dozen homes:** R. Bronen and F. Chapin III, "Adaptive Governance and Institutional Strategies for Climate-Induced Community Relocations in Alaska," *Proceedings of the National Academy of Sciences* 110, no. 23 (June 4, 2013), https://www.pnas.org/content/110/23/9320.

161   **In 2002, the village again voted:** "Shishmaref Relocation and Collocation Study," U.S. Army Corps of Engineers, Alaska District, December 2004, https://web.law.columbia.edu/sites/default/files/microsites/climate-change/files/Arctic-Resources/Relocation-Plans/USACE%20relocation%20plan_shishmaref[1].pdf.

161   **Instead, officials funneled $27 million:** Christopher Mele and Daniel Victor, "Reeling from Effects of Climate Change, Alaskan Village Votes to Relocate," *New York Times,* August 9, 2016, https://www.nytimes.com/2016/08/20/us/shishmaref-alaska-elocate-vote-climate-change.html.

161   **Pulitzer Prize winner Elizabeth Kolbert:** Elizabeth Kolbert, *Field Notes from a Catastrophe* (London: Bloomsbury, 2006).

162   **Like their counterparts in Shishmaref:** "Climate Change in Kivalina, Alaska," Alaska Native Tribal Health Consortium, January 2011, https://www.cidrap.umn.edu/sites/default/files/public/php/26952/Climate%20Change%20HIA%20Report_Kivalina.pdf.

162   **All told, more than two hundred:** United States Government Accountability Office, *Alaska Native Villages: Limited Progress Has Been Made on Relocating Villages Threatened by Flooding and Erosion,* GAO 09-551 (Washington, DC: Government Accountability Office, 2009), https://www.gao.gov/assets/300/290468.pdf.

162   **Svante Arrhenius, the nineteenth-century:** S. Arrhenius, "On the Influence of Carbonic Acid in the Air upon the Temperature of the Ground," *Philosophical Magazine and Journal of Science,* series 5, vol. 41 (April 1896): 237–76, https://www.rsc.org/images/Arrhenius1896_tcm18-173546.pdf.

163   **Zooplankton come in a wide variety:** "Climate Drives Change in an Arctic Food Web," NOAA Fisheries, May 7, 2019, https://www.fisheries.noaa.gov/feature-story/climate-drives-change-arctic-food-web.

164 **The most recent research:** P. Molnár, C. Bitz, M. Holland, J. Kay, S. Penk, and S. Amstrup, "Fasting Season Length Sets Temporal Limits for Global Polar Bear Persistence," *Nature Climate Change* 10 (July 20, 2020): 732–38, https://www.nature.com/articles/s41558-020-0818-9.

165 **In the Northern Hemisphere:** Dennis Mersereau, "What Is the Jet Stream, and How Does It Work?" *Mental Floss,* August 28, 2017, https://www.mentalfloss.com/article/503501/what-jet-stream-and-how-does-it-work.

165 **A weakened jet stream:** M. Mann, S. Rahmstorf, K. Kornhuber, B. Steinman, S. Miller, and D. Coumou, "Influence of Anthropogenic Climate Change on Planetary Wave Resonance and Extreme Weather Events," *Scientific Reports* 7, no. 45242 (March 27, 2017), https://www.nature.com/articles/srep45242.

165 **some studies have shown links:** Bob Berwyn, "Wobbly Jet Stream Is Sending the Melting Arctic into 'Uncharted Territory,'" InsideClimate News, June 9, 2016, https://insideclimatenews.org/news/08062016/greenland-arctic-record-melt-jet-stream-wobbly-global-warming-climate-change.

165 **Without the implementation of costly adaptation:** "The Ocean and Cryosphere in a Changing Climate, Summary for Policymakers," Intergovernmental Panel on Climate Change, United Nations, September 24, 2019, https://report.ipcc.ch/srocc/pdf/SROCC_SPM_Approved.pdf.

166 **A 2019 study estimated:** "Flooded Future: Global Vulnerability to Sea Level Rise Worse Than Previously Understood," Climate Central, October 29, 2019, https://www.climatecentral.org/pdfs/2019CoastalDEMReport.pdf.

167 **The following year, Sinnok's uncle Norman:** "Man Dies After Snowmachine Falls Through Shishmaref Ice," *Anchorage Daily News,* June 3, 2007, https://www.akfatal.net/Kokeok%2006-02-07.htm.

167 **Today, a white cross that marks:** John D. Sutter, "Tragedy of a Village Built on Ice," CNN, March 29, 2017, https://www.cnn.com/2017/03/29/us/sutter-shishmaref-esau-tragedy/index.html.

168 **When the federal government began building:** Robin Bronen, "Climate-Induced Displacement of Alaska Native Communi-

ties," Brookings-LSE Project on Internal Displacement, Brookings Institution, January 30, 2013, https://www.brookings.edu/wp-content/uploads/2016/06/30-climate-alaska-bronen-paper.pdf.

168 **There he met Itinterunga Rae Bainteiti:** "Esau and Rae," conversation filmed at 2015 United Nations Climate Change Conference, YouTube video, 3:17, uploaded by Benj Brooking, December 12, 2015, https://www.youtube.com/watch?v=fT45DUEPYXc.

169 **President Obama hailed the agreement:** Barack Obama, "Remarks by the President on the Paris Agreement," National Archives and Records Administration, accessed May 10, 2020, https://obamawhitehouse.archives.gov/the-press-office/2016/10/05/remarks-president-paris-agreement.

169 **At the heart of the lawsuit:** Sinnok v. Alaska, 3AN-17-09910 CI (Alaska Super. Ct. 2017), http://blogs2.law.columbia.edu/climate-change-litigation/wp-content/uploads/sites/16/case-documents/2017/20171027_docket-3AN-17-09910_complaint.pdf.

170 **In 2008, the villagers of Kivalina:** Lawrence Hurley, "Supreme Court Declines to Hear Alaska Climate Change Case," Reuters, May 20, 2013, https://www.reuters.com/article/us-usa-court-climate/supreme-court-declines-to-hear-alaska-climate-change-case-idUSBRE94J0FQ20130520.

170 **In 2015, Kelsey Juliana:** "Federal Youth Climate Case—Juliana v. U.S.," Crag Law Center, accessed May 10, 2020, https://crag.org/federal-youth-climate-case-juliana-v-u-s/.

171 **"In these proceedings, the government accepts":** Juliana v. United States, 947 F.3d 1159 (9th Cir., 2020), http://cdn.ca9.uscourts.gov/datastore/opinions/2020/01/17/18-36082.pdf.

Chapter 17: Home, for Now

173 **In a nine-minute video:** "Nora's Second Chance," YouTube video, 8:40, Oregon Zoo, September 29, 2017, https://www.youtube.com/watch?v=lWZBi2i0QUo.

175 **She'd loved animals ever since:** Nour Habib, "What Are You . . . ? With Kaleigh Jablonski, Tulsa Zoo Animal Trainer,"

Tulsa World, July 12, 2012, https://www.tulsaworld.com/archive/ what-are-you-with-kaleigh-jablonski-tulsa-zoo-animal-trainer/ article_e49c9473-0079-5f99-a832-371d887246a0.html.

176 **Around 8:15 A.M.:** "Polar Bear Update! Hope and Nora!" Facebook Live video, 11:15, Utah's Hogle Zoo, October 5, 2017, https://www.facebook.com/watch/live/?v=10159419889760173 &ref=watch_permalink.

178 **In another video update:** "It's a Hope and Nora Update Plus a Note About Climate Change—LIVE #utahclimateweek," Facebook Live video, 16:07, Utah's Hogle Zoo, October 12, 2017, https://www.facebook.com/watch/live/?v=10159450016625173 &ref=watch_permalink.

179 **For example, there's a long-running:** "Global Warming vs Climate Change," Skeptical Science, accessed May 10, 2020, https:// skepticalscience.com/climate-change-global-warming.htm.

179 **One of the foremost researchers:** N. Oreskes, "The Scientific Consensus on Climate Change," *Science* 306, no. 5702 (December 3, 2004): 1686, https://science.sciencemag.org/content/306/ 5702/1686.

181 **A 2009 survey found:** J. Cook, S. van der Linden, E Maibach, and S. Lewandowsky, *The Consensus Handbook,* George Mason Center for Climate Change Communication, 2018, http://www .climatechangecommunication.org/all/consensus-handbook/.

181 **"Emphasize the uncertainty in scientific conclusions":** "1988 Exxon Memo on the Greenhouse Effect," Climate Files, accessed May 10, 2020, http://www.climatefiles.com/exxonmobil/566/.

181 **An analysis of news programs:** M. Boykoff, "Lost in Translation? United States Television News Coverage of Anthropogenic Climate Change, 1995–2004," *Climatic Change* 86 (January 2008), https://link.springer.com/article/10.1007/s10584-007-9299-3.

181 **In a 2017 interview:** Coral Davenport, "With Climate Science on the March, an Isolated Trump Hunkers Down," *New York Times,* February 28, 2019, https://www.nytimes.com/2019/02/28/ climate/trump-climate-science.html.

182 **They point to the fact that:** "How Is Today's Warming Different from the Past?," NASA Earth Observatory, June 3, 2010, https:// earthobservatory.nasa.gov/features/GlobalWarming/page3.php.

182 **Crockford grew up in Canada:** Susan Crockford, "On Being a Polar Bear Expert, Among Other Things," *Polar Bear Science* (blog), March 12, 2015, https://polarbearscience.com/2015/03/12/on-being-a-polar-bear-expert-among-other-things/.

182 **Crockford attended the University:** Susan Crockford biography, Pacific IDentifications website via Wayback Machine, accessed May 10, 2020, https://web.archive.org/web/20171226030457/http://pacificid.com/pages-added/about-pacific-id.php.

183 **"I've had quite enough":** Susan Crockford, "Cooling the Polar Bear Spin," *Polar Bear Science* (blog), July 26, 2012, https://polarbearscience.com/2012/07/26/cooling-the-polar-bear-spin/#more-98.

183 **She argues that the number:** "Polar Bear Status Table," IUCN/SSC Polar Bear Specialist Group, September 2019, accessed May 9, 2020, http://pbsg.npolar.no/export/sites/pbsg/en/docs/2019-PBSG-StatusTable.pdf.

183 **Though nearly every organization:** "Global Polar Bear Population Estimates," IUCN/SSC Polar Bear Specialist Group, July 11, 2014, accessed May 9, 2020, http://pbsg.npolar.no/en/status/pb-global-estimate.html.

183 **Crockford made her own "best guess":** Susan Crockford, "Latest Global Polar Bear Abundance 'Best Guess' Estimate Is 39,000 (26,000–58,000)," *Polar Bear Science* (blog), March 26, 2019, https://polarbearscience.com/2019/03/26/latest-global-polar-bear-abundance-best-guess-estimate-is-39000-26000-58000/.

184 **"I took photographs, and Paul recorded":** Cristina Mittermeier, "Starving-Polar-Bear Photographer Recalls What Went Wrong," *National Geographic,* August 2018, https://www.nationalgeographic.com/magazine/2018/08/explore-through-the-lens-starving-polar-bear-photo/#close.

184 ***The Washington Post* called the heartbreaking:** Eli Rosenberg, "'We Stood There Crying': Emaciated Polar Bear Seen in 'Gut-Wrenching' Video and Photos," *Washington Post,* December 9, 2017, https://www.washingtonpost.com/news/animalia/wp/2017/12/09/we-stood-there-crying-the-story-behind-the-emotional-video-of-a-starving-polar-bear/.

184 **"One starving bear is not evidence":** Susan Crockford, "One Starving Bear Is Not Evidence of Climate Change, Despite Grue-

some Photos," *Polar Bear Science* (blog), December 9, 2017, https://polarbearscience.com/2017/12/09/one-starving-bear-is -not-evidence-of-climate-change-despite-gruesome-photos/ #more-113805.

185 **"Perhaps we made a mistake":** Mittermeier, "Starving-Polar-Bear Photographer."

185 **Through a process of citation:** J. Harvey, D. van den Berg, J. Ellers, R. Kampen, T. Crowther, P. Roessingh, B. Verheggen, R. Nuijten, E. Post, S. Lewandowsky, I. Stirling, M. Balgopal, S. Amstrup, and M. Mann, "Internet Blogs, Polar Bears, and Climate-Change Denial by Proxy," *BioScience* 68, no. 4 (April 2018): 281–87, https://academic.oup.com/bioscience/article/ 68/4/281/4644513.

186 **Crockford responded in a series:** Susan Crockford, "Bioscience Article Is Academic Rape: An Assertion of Power and Intimidation," *Polar Bear Science* (blog), December 2, 2017, https://polar bearscience.com/2017/12/02/bioscience-article-is-academic -rape-an-assertion-of-power-and-intimidation/.

186 **demanding a retraction:** Susan Crockford, "Retraction Request to Bioscience: FOIA Emails Document Another Harsh Criticism of Amstrup's 2007 Polar Bear Model," *Polar Bear Science* (blog), December 5, 2017, https://polarbearscience.com/2017/12/05/ retraction-request-to-bioscience-foia-emails-document-another -harsh-criticism-of-amstrups-2007-polar-bear-model/.

186 **In 2019, Crockford's contract at:** Susan Crockford, "UVic Bows to Outside Pressure and Rescinds My Adjunct Professor Status," *Polar Bear Science* (blog), October 16, 2019, https://polarbear science.com/2019/10/16/uvic-bows-to-outside-pressure-and -rescinds-my-adjunct-professor-status/.

186 **The university said it had nothing:** Brishti Basu, "Climate Change Denier Loses Adjunct Professor Status at University of Victoria (Updated)," *Victoria Buzz,* October 28, 2019, https:// www.victoriabuzz.com/2019/10/climate-change-denier-loses -adjunct-professor-status-at-university-of-victoria/.

Chapter 18: Broken

191 **Shannon Morarity, the Nora Mom:** Columbus Zoo and Aquarium, "Some of you might remember me . . . ," Facebook, June 11,

2019, https://www.facebook.com/columbuszoo/photos/a.1051 62722105/10156559611037106/?type=3&theater.

193 **Winter got off to a slow start:** "Baked Alaska and 2017 in Review," National Snow & Ice Data Center, January 5, 2017, http://nsidc.org/arcticseaicenews/2018/02/.

193 **In late February:** Diana Haecker, "Open Water Wreaks Havoc at Little Diomede," *Nome Nugget,* February 2, 2018, http://www.nomenugget.com/news/open-water-wreaks-havoc-little-diomede.

193 **A National Weather Service meteorologist:** Alex DeMarban, "Storms Pummel Bering Sea Islands After 'Crazy' Ice Melt-Off," *Anchorage Daily News,* February 27, 2018, https://www.adn.com/alaska-news/rural-alaska/2018/02/27/warm-storms-pummel-bering-sea-leading-to-crazy-ice-melt-off/.

194 **The story was similar farther south:** International Arctic Research Center, "In the Coastal Communities Near the Bering Strait, a Winter Unlike the Rest," NOAA Climate.gov, April 16, 2018, https://www.climate.gov/news-features/features/coastal-communities-near-bering-strait-winter-unlike-rest.

194 **In February, the closest weather station:** Michon Scott, "February 2018 Heatwave Across the Far North," NOAA Climate.gov, March 20, 2018, https://www.climate.gov/news-features/event-tracker/february-2018-heatwave-across-far-north.

194 **At one point, the temperature:** Kendra Pierre-Louis, "Europe Was Colder Than the North Pole This Week. How Could That Be?," *New York Times,* March 1, 2018, https://www.nytimes.com/2018/03/01/climate/polar-vortex-europe-cold.html.

194 **In July, several Scandinavian countries:** Jason Samenow, "Scorching Scandinavia: Record-Breaking Heat Hits Norway, Finland and Sweden," *Washington Post,* July 18, 2018, https://www.washingtonpost.com/news/capital-weather-gang/wp/2018/07/17/scorching-scandinavia-record-breaking-heat-hits-norway-finland-and-sweden/.

195 **In Sweden, the country's tallest peak:** Christina Anderson, "Sweden's Tallest Peak Shrinks in Record Heat," *New York Times,* August 2, 2018, https://www.nytimes.com/2018/08/02/world/europe/sweden-kebnekaise-heat-wave.html?module=inline.

195 **The heat was not confined:** Jason Samenow, "Red-Hot Planet: All-Time Heat Records Have Been Set All Over the World During

the Past Week," *Washington Post,* July 5, 2018, https://www
.washingtonpost.com/news/capital-weather-gang/wp/2018/07/
03/hot-planet-all-time-heat-records-have-been-set-all-over-the
-world-in-last-week/.

195 **Wildfires, in most cases started:** Niki Kitsantonis, Richard
Pérez-Peña, and Russell Goldman, "In Greece, Wildfires Kill
Dozens, Driving Some into the Sea," *New York Times,* July
24, 2018, https://www.nytimes.com/2018/07/24/world/europe/
greece-fire-deaths.html.

195 **Britain and Sweden both experienced:** Jonathan Watts, "The
Swedish Town on the Frontline of the Arctic wildfires," *Guardian,* July 30, 2018, https://www.theguardian.com/world/2018/
jul/30/the-swedish-town-on-the-frontline-of-the-arctic
-wildfires.

195 **After years of drought:** "Camp Fire: Green Sheet," California
Department of Forestry and Fire Protection, accessed May 10,
2020, https://assets.documentcloud.org/documents/5628194/18
-CA-BTU-016737-Camp-Green-Sheet.pdf.

195 **It was the deadliest:** Priyanka Boghani, "Camp Fire: By the
Numbers," *Frontline,* PBS, October 29, 2019, https://www.pbs
.org/wgbh/frontline/article/camp-fire-by-the-numbers/.

196 **"Climate change has rendered the term":** "2018 Strategic Fire
Plan for California," California Department of Forestry and Fire
Protection, August 22, 2018, https://osfm.fire.ca.gov/media/
5590/2018-strategic-fire-plan-approved-08_22_18.pdf.

196 **EPA administrator, Scott Pruitt:** "EPA Withdraws Information
Request for the Oil and Gas Industry," news release, Environmental Protection Agency via Wayback Machine, February 2,
2017, accessed May 10, 2020, https://web.archive.org/web/
20180124201514/https://www.epa.gov/newsreleases/epa
-withdraws-information-request-oil-and-gas-industry.

196 **The Obama administration had instituted:** John Wihbey,
"Understanding the Social Cost of Carbon—and Connecting It
to Our Lives," Yale Climate Connections, February 12, 2015,
https://www.yaleclimateconnections.org/2015/02/
understanding-the-social-cost-of-carbon-and-connecting-it-to
-our-lives/.

196 **Trump rescinded it:** Donald J. Trump, Exec. Order No. 13783,
Promoting Energy Independence and Economic Growth, 82

Fed. Reg. 16093 (March 28, 2017), https://www.whitehouse.gov/presidential-actions/presidential-executive-order-promoting-energy-independence-economic-growth/.

196   **Trump lifted it:** Eric Lipton and Barry Meier, "Under Trump, Coal Mining Gets New Life on U.S. Lands," *New York Times,* August 6, 2017, https://www.nytimes.com/2017/08/06/us/politics/under-trump-coal-mining-gets-new-life-on-us-lands.html.

197   **Trump rescinded that, too:** Diana Haecker, "Trump Issues Executive Order Revoking Northern Bering Sea Protection and Tribal Participation," *Nome Nugget,* May 5, 2017, http://www.nomenugget.com/news/trump-issues-executive-order-revoking-northern-bering-sea-protection-and-tribal-participation.

197   **He withdrew an order:** Rob Hotakainen, "NPS Chief Scraps Climate-Focused Order," E&E News, August 31, 2017, https://www.eenews.net/stories/1060059511.

197   **In October of that year:** "Summary for Policymakers of IPCC Special Report on Global Warming of 1.5°C Approved by Governments," Intergovernmental Panel on Climate Change, accessed May 10, 2020, https://www.ipcc.ch/2018/10/08/summary-for-policymakers-of-ipcc-special-report-on-global-warming-of-1-5c-approved-by-governments/.

197   **That half degree would be crucial:** Kelly Levin, "8 Things You Need to Know About the IPCC 1.5°C Report," World Resources Institute, October 7, 2018, https://www.wri.org/blog/2018/10/8-things-you-need-know-about-ipcc-15-c-report.

197   **Six weeks later, the U.S. government's:** *Impacts, Risks, and Adaptation in the United States: Fourth National Climate Assessment,"* U.S. Global Change Research Program, Washington, DC, November 2018, accessed May 10, 2020, https://nca2018.globalchange.gov/chapter/front-matter-about/.

Chapter 19: A Risky Repair

202   **Doctors have experimented with:** David Seligson, "History of Intramedullary Nailing," in Pol M. Rommens and Martin H. Hessmann (Eds.), *Intramedullary Nailing: A Comprehensive Guide* (London: Springer, 2015), https://www.researchgate.net/publication/302529643_History_of_Intramedullary_Nailing.

203 **He met another surgeon there:** "Unique Patient, Unique Challenge," AO Vet, February 14, 2018, https://aovet.aofoundation .org/en/about-aovet/news/news-2018/polar-bear-amy-kapatkin.

203 **He'd consider it, he told Crook:** "Texas A&M Professors Perform First Humerus Repair on Polar Bear," College of Veterinary Medicine & Biological Sciences, Texas A&M University, March 4, 2019,   https://vetmed.tamu.edu/news/press-releases/texas-am -professors-perform-first-humerus-repair-on-polar-bear/.

207 **Margolin identifies as:** "Jamie Margolin, 17," International Congress of Youth Voices, accessed May 11, 2020, https://www .internationalcongressofyouthvoices.com/jamie-margolin.

207 **In 2018, like Esau Sinnok:** John Ryan, "13 Kids Sue Washington State for Life, Liberty and a Livable Climate," KUOW, February 20, 2018, http://archive.kuow.org/post/13-kids-sue -washington-state-life-liberty-and-livable-climate.

207 **That summer, while Nora was playing:** "Our Actions," Zero Hour, accessed May 11, 2020, http://thisiszerohour.org/our -actions/.

207 **They called on Congress:** "Dear Elected Officials," Zero Hour, accessed May 11, 2020, http://thisiszerohour.org/files/zh -platform-politicians-web.pdf.

208 **Two days after delivering those demands:** Kristen Doerer, "Youth Climate Change Activists Marched on Washington, D.C.," *Teen Vogue,* July 22, 2018, https://www.teenvogue.com/ story/youth-climate-change-activists-marched-washington-dc.

208 **On August 20, she posted:** Greta Thunberg (@gretathunberg), photo of first school strike, Instagram, August 20, 2018, https:// www.instagram.com/p/BmsTxPPl0qW/?hl=en.

208 **Within a week, thirty-five people:** Amelia Tait, "Greta Thunberg: How One Teenager Became the Voice of the Planet," *Wired,* June 6, 2019, https://www.wired.co.uk/article/greta-thunberg -climate-crisis.

209 **After the Democrats won a majority:** David Roberts, "Alexandria Ocasio-Cortez Is Already Pressuring Nancy Pelosi on Climate Change," *Vox,* November 15, 2018, https://www.vox.com/ energy-and-environment/2018/11/14/18094452/alexandria -ocasio-cortez-nancy-pelosi-protest-climate-change-2020.

Chapter 20: Nora's Keepers

212 **The next day, Nora was awake:** "Nora Update!!," Facebook Live video, 16:55, Utah's Hogle Zoo, February 5, 2019, https://www .facebook.com/watch/live/?v=599343567146085&ref=watch _permalink.

213 **The vets had prescribed trazodone:** "NORA Update!! Let's Peek In on the Sweet Girl!," Facebook Live video, 7:48, Utah's Hogle Zoo, February 19, 2019, https://www.facebook.com/ watch/?v=391573158336630.

213 **At ten weeks, they sedated Nora:** "Nora Checkup," Facebook Live video, 12:19, Utah's Hogle Zoo, April 19, 2019, https:// www.facebook.com/watch/?v=3252635458096019.

214 **As soon as she saw:** "Nora Takes a Dip! This is Nora's first swim since having surgery in February . . . ," Facebook Live video, 0:43, Utah's Hogle Zoo, May 28, 2019, https://www.facebook.com/ watch/?v=313178716250840.

214 **It was troublingly similar to the previous:** "Unprecedented 2018 Bering Sea Ice Loss Repeated in 2019," NOAA, August 14, 2019, https://www.noaa.gov/stories/unprecedented-2018 -bering-sea-ice-loss-repeated-in-2019.

214 **Frozen rivers in the southeast:** Yereth Rosen, "Record-Early Alaska River Breakups Are Part of a Long-Term Warming Trend," Arctic Today, April 15, 2019, https://www.arctictoday.com/ record-early-alaska-river-breakups-are-part-of-a-long-term -warming-trend/.

214 **The Trump administration mandated:** Coral Davenport and Mark Landler, "Trump Administration Hardens Its Attack on Climate Science," *New York Times,* May 27, 2019, https://www .nytimes.com/2019/05/27/us/politics/trump-climate-science .html.

215 **CNN hosted a seven-hour climate crisis:** Meg Wagner, Dan Merica, Gregory Krieg, and Eric Bradner, "CNN's Climate Crisis Town Hall," CNN, September 5, 2019, https://www.cnn.com/ politics/live-news/climate-crisis-town-hall-august-2019/index .html.

215 **Greta Thunberg traveled to New York:** "Greta Thunberg Will Sail Across the Atlantic on a Zero-Emissions Yacht for the UN

Climate Summit," CNN, August 18, 2019, https://www.cnn
.com/2019/07/29/europe/greta-thunberg-sailboat-scli-intl/
index.html.

215 **"The fact that you are staring"**: *Voices Leading the Next Generation on the Global Climate Crisis: House Hearing Before the Subcommittee on Europe, Eurasia, Energy, and the Environment of the Committee on Foreign Affairs*, 116th Cong. (2019) (statement of Jamie Margolin), https://docs.house.gov/meetings/FA/FA14/20190918/109951/HHRG-116-FA14-Wstate-MargolinJ-20190918.pdf.

215 **Both were the subject of online:** Zahra Hirji, "Teenage Girls Are Leading the Climate Movement—and Getting Attacked for It," *BuzzFeed News*, September 25, 2019, https://www.buzzfeednews .com/article/zahrahirji/greta-thunberg-climate-teen-activist -harassment.

216 **There were more than eight hundred rallies:** Scott Neuman and Bill Chappell, "Young People Lead Millions to Protest Global Inaction on Climate Change," NPR, September 20, 2019, https://www.npr.org/2019/09/20/762629200/mass-protests-in -australia-kick-off-global-climate-strike-ahead-of-u-n-summit.

216 **A small group gathered in Brevig:** "Youth Climate Strike," Alaska Center Education Fund, accessed May 11, 2020, https://akcentereducationfund.org/ayea/youth-climate-strikes/.

216 **Hundreds marched in Columbus:** Erica Thompson, "Hundreds of Columbus Students Cut Class to Participate in Global Climate Strike," *Columbus Dispatch*, September 20, 2019, https://www.dispatch.com/news/20190920/hundreds-of-columbus -students-cut-class-to-participate-in-global-climate-strike.

216 **In Salt Lake City, they marched:** Brian Maffly, " 'Winter Is Not Coming'—Hundreds of Young Utahns Protest, Demand Action on Climate Change," *Salt Lake Tribune*, September 20, 2019, https://www.sltrib.com/news/environment/2019/09/20/utah -students-join-global/.

217 **The groundskeepers at the zoo:** "How Is Nora Doing? Let's Find Out!," Facebook Live video, 13:19, Utah's Hogle Zoo, June 25, 2019, https://www.facebook.com/watch/live/?v=22389 23409560202&ref=watch_permalink.

221 **In the world of climate deniers:** Anthony Watts, "Halloween Climate Scare #5: We Have 12 Years to Save the World," *Watts Up with That?* (blog), October 27, 2019, https://wattsupwiththat .com/2019/10/27/halloweenv-climate-scare-5-we-have-12-years -to-save-the-world/.

Index

About the Author

KALE WILLIAMS is a reporter at *The Oregonian*/OregonLive, where he covers science and the environment. A native of the Bay Area, he previously reported for the *San Francisco Chronicle*. He shares a home with his wife, Rebecca; his two dogs, Goose and Beans; his cat, Torta; and his stepcat, Lucas.

kale-williams.com
Twitter: @sfkale

About the Type

This book was set in Legacy, a typeface family designed by Ronald Arnholm (b. 1939) and issued in digital form by ITC in 1992. Both its serifed and unserifed versions are based on an original type created by the French punch cutter Nicholas Jenson in the late fifteenth century. While Legacy tends to differ from Jenson's original in its proportions, it maintains much of the latter's characteristic modulations in stroke.